WHAT DAD DID in the WAR

S/SGT William N. Finney [c].

WHAT DAD DID in the WAR

A *Colorized* Story of the Italian Campaign

★ ★ ★

WILLIAM P. FINNEY

ISBN: 978-1-7363893-0-0

LCCN: 2024916831

Publisher's Cataloging-in-Publication:

Names: Finney, William P., author.
Title: What Dad Did in the War: a Colorized Story of the Italian Campaign/ William P. Finney.
Description: Clearwater, FL : Dayminimis Press, [2024] | Includes bibliographical references and index.
Identifiers: ISBN: 978-1-7363893-0-0 (hardcover) | LCCN: 2024916831
Subjects: LCSH: Finney, William N. | Depressions–1929–Wisconsin–Biography. | Children–Wisconsin–History–20th century–Biography. | Soldiers–United States–Biography. | World War, 1939-1945–Wisconsin. | World War, 1939-1945–Africa, North. | World War, 1939-1945–Mediterranean Region. | World War, 1939–1945–Campaigns–Italy. | World War, 1939- 1945–Pictorial works. | BISAC: HISTORY / Wars & Conflicts / World War II / Mediterranean Theater. | BIOGRAPHY & AUTOBIOGRAPHY / Military.
Classification: LCC: D763.I8 .F56 2024 | DDC: 940.54/215092–dc23

DAYMINIMIS PRESS
Clearwater, FL 33761

Design by Marina Drukman

Printed in the United States of America

To my beloved wife Cheryl, who has
tolerated my eccentricities all these years,
to my children, John, Colleen and James,
to my mom and step-dad, Marge and Ron Marten,
to my sisters, Mary Ponto and Patty Yttre,
to Bonnie L.,
to all those veterans whose records are lost forever,
and,
of course,
to Dad.

CONTENTS

INTRODUCTION

In May of 2016, eleven months after my request, I received a copy of my father's reconstructed, *partially-burned*, World War II Army records from the National Personnel Records Center in St. Louis, Missouri. His records, along with the military records of millions of other soldiers, were believed lost in a devastating fire at the Records Center that occurred July 12, 1973, and burned for twenty-two hours. At the time, the Records Center had no smoke detectors or sprinklers. My dad had tried to get copies of his military records and was told they were lost in the fire. With Dad having died in 1995, aside from what I could remember him telling me about his exploits in North Africa, Corsica, and Italy, I lost hope of ever finding out exactly where he went or what he did during his military service in World War II. I regretted never having put him in front of a tape recorder and asking him what he could recall.

National Personnel Records Center fire in suburban St. Louis, 1973. (*St. Louis Post-Dispatch*).[5]

In June of 2015, I learned the National Personnel Records Center and the National Archives and Records Administration had managed to salvage 20 percent of the lost military records of the World

War II Army vets, which, for a fee, could be reconstructed by archivists—*and Dad's part of the alphabet was included.*

Eagerly, I applied for a copy of Dad's records.

From what was salvaged, what I remember Dad telling me, the records, documents, and photos Dad had kept, his online draft records, newspaper reports, local and family history, correspondence, and extensive research—including a review of unit "morning reports" at the NPRC in St. Louis—I determined to put down, as best as I could, an account of Dad's "Greatest Generation" wartime experience. In addition, I felt compelled to tell the story of Dad's poignant childhood during the Great Depression, growing up with his divorced, bartender uncle in blue-collar West Allis, Wisconsin—as well as an account of significant events toward the end of the War in Italy while Dad was there—including the stories of the capture and death of Mussolini, and the final surrender.

Dad kept no diary while in service, and I have no wartime letters from or to him. Re-creating Dad's story was only possible because of the photos and documents he kept, and the good fortune of recovering his military records, once believed burned to ashes in the fire.

How sad, that the records, and therefore the histories, of some 80 percent of the World War II Army veterans have been *lost forever.* In 1995, when Dad died, the World Wide Web was in its infancy. To tell Dad's story *today,* I wanted to take advantage of the digital revolution—not only for research into the past, but also for *imaging* in this present visual age.

To illustrate Dad's story with his photos and those of his time, I opted to take advantage of *artificial intelligence colorization.* I appreciate that colorization of old black-and-white photos is controversial and has many skeptics. I know color is highly subjective—my doctor told me my old corneas have actually yellowed with age—and colorization frequently is inaccurate and, for many, even a distraction. However, we do see in color and are used to adjusting to color changes in the many lighting situations we encounter every day. I admit colorization has many shortcomings and does not work equally well in all cases. Nevertheless, despite my obvious reservations, and after seeing results that ranged from what I thought were spectacular to so-so, I ultimately decided colorization added to the reality of Dad's story and made it more interesting. I hope you agree. For the reader's benefit, in the text beneath each black-and-white photo that has been colorized, I have added a small superscript [c] (for *colorized*).

My initial goal of retracing my father's steps led me on a journey I never anticipated as I did my best to decipher the bureaucratic entries in Dad's military records and learn about the Italian conflict in general. War correspondent Ernie Pyle said Italy was World War II's "forgotten front." I am a post-war "baby-boomer." I had nothing to forget. For me, this was a journey of discovery—though, sadly, for the descendants of the vast majority of World War II Army vets, whose records are lost, a journey that cannot be undertaken. Fortunate to have Dad's records, such as they were, I felt almost *obligated* to tell his truly American story—and I learned so much not only about him, but also about our country, about our military, and about Dad's heavy-industry Wisconsin hometown in time of war. To understand what Dad did, and to trace his movements, I wanted to find out the why of things, which led me often to look at the bigger picture—involving much Dad himself never knew.

The War has provided a rich tapestry of historical material. Generals and many ordinary soldiers have written memoirs; the historians have written histories; and the sociologists have written of the impact of the War. Today, as noted above, the internet provides resources unimaginable in the past, many in Italian, with which I have some familiarity.

This is a micro/macro tale of a "GI Joe," a member of the "Greatest Generation," whose young life was molded by the Great Depression of the 1930s and World War II—and also, about the fascinating story of the War in Italy in which he was a participant.

From time to time Dad told me about his experiences in the Depression and the War. Many of those stories are in this narrative.

I only wish he were still around to share more of them with us. Dad's story, while in ways unique, is mainly that of an average American soldier in World War II. But when the War ended, those average soldiers, like Dad, made history as they came back home *by the millions* from exotic places they previously only imagined, with experiences they never anticipated, some they could never forget—some they wanted to forget. Those vets, of all races, religions, and ethnic backgrounds, having lived with the stress of combat in foreign lands, came home, got jobs, married, raised families, took advantage of the GI Bill—and dramatically changed our whole society. Yet, for all their postwar accomplishments, most vets, if asked, would admit their participation in World War II was the most significant time in their lives.

That is how it was for Dad.

The Wisconsin Veterans Museum has a saying: *Every veteran is a story.*

This one is Dad's.

1

BACKGROUND

My father, **William Norbert Finney**, was born in West Allis, Wisconsin, September 11, 1924. To my surprise, I learned *that is not what his military records say*—but I'll get to that later.

"I was a kid of the streets," he used to tell me of his unfortunate, sometimes tragic, childhood, growing up in blue-collar West Allis, suburban Milwaukee, during the hard times of the Great Depression of the 1930s. A story unto itself, the family hardships and misfortunes—involving abandonment, ill-fated "May-December" marriages, and untimely deaths—began before my father was born, and involved several veterans of other wars. It began with Dad's grandmother.

Dad's grandmother, my great-grandmother, Catherine Ellen Lark, called "Kittie," the person who probably had the second-greatest influence on Dad growing up, was Chicago-Irish—or at least *half-so*. Destined for heartache, she was born in 1870, the daughter of Civil War veteran William Lark, a Chicago house painter. In the tragic year 1883, when Catherine was the tender age of only thirteen, death took her mother, Bridget "Lucy" Reynolds Lark, only forty-two, *and* her grandmother, Ann Reynolds, seventy—as well as an infant sister born that year. Younger brother Willie died of typhoid at age eighteen in 1891.

By 1899, and probably sooner, she had followed her veteran father to the Milwaukee area and was living as a "widow" with three children, all born in Chicago, all surnamed Boulton: Marie Catherine, called "Mae," born in 1892; Francis R., born in 1895; and William Leroy, born in 1896. I have not found any marriage license for Dad's grandmother and anyone named Boulton (or *Bolton*, as the name was sometimes spelled).

Chicago city directories paint a peculiar picture I cannot understand. In 1893, the year after her first child Mae was born, Catherine was listed as "Katherine [sic] E. Boulton, widow of *Frank*." The birth and baptismal records for Mae list the father as an electrician named Francis Boulton. As noted above, the supposed widow had two more children with the surname Boulton three and four years after that.

How was that possible for the *widow* Boulton?

In 1896 her brother George Lark was living with an electrician named Frank R. Boulton. (i.e., her "dead husband" was still *alive*). In 1897, after the births of her second and third children, she was in the Chicago directory as "Catherine Boulton, widow of *Robert*." Baptismal records for the second and third child list the father as Robert Joseph Bolton—not Francis or Frank Boulton.

We know Catherine was living in Wisconsin in 1899 because her second-born, Francis R. Boulton, died in Milwaukee County on February 9 of that year of bronchitis at only three years of age. On the death certificate, as on his baptismal record, the child's father was listed as "*Robert J. Boulton*,"—not Frank as on his birth certificate.

So, when Catherine "Boulton" and her three children came to Wisconsin, was she once a widow (Frank *or* Robert, or were they the *same man*?), twice a widow (Frank *and* Robert as listed in Chicago directories?)—or was she a widow at all? As we shall see, "widow" at that time was a convenient term for a woman with children seeking government welfare—and no husband around.

Like Catherine's Irish-born mother, Bridget "Lucy" (Reynolds) Lark, and Irish-born grandparents, Ann and Michael Reynolds, young Francis, who died in Wisconsin in 1899, was buried at Calvary Catholic Cemetery in the northern Chicago suburb of Evanston, Illinois.

By the 1900 census, Dad's grandmother Catherine—as Catherine Boulton, "a widow"—and her two surviving children (Mae and William) were living with a telephone lineman named John Samuel Spearman near the Milwaukee Veterans Home in unincorporated Wauwatosa Township. The enumeration day for the census was June 22. The following day, Catherine and John, were married in Milwaukee. Catherine gave her name as "Catherine E. Lark," not Boulton, as on the census. The marriage, supposedly performed by a Catholic priest, was nevertheless designated a "civil ceremony" on the marriage license.

Catherine's widowed father William Lark, who claimed to have been born in New York—but was actually born (as *Wilhelm Lerch*) in Germany—had been the *first* of Dad's family to move from Chicago to the Milwaukee area.

In 1895, he was admitted to what was officially called the Northwest Branch of the U.S. National Home for Disabled Volunteer Soldiers. In

The Lark Family in Chicago, dated 1885. Catherine, William, Willie, Bridget, and George. (*This is a doctored photo, common at the time, since Bridget, or "Lucy," the mother, had died in 1883.*)

1866, Milwaukee had been chosen as the site of one of the three original Homes for disabled Civil War veterans. The 375-acre scenic Northwest Branch property included *four* small lakes, a Gothic Revival "Old Main" building, as well as a hospital, chapel, library, theater, and cemetery, among other facilities.

The Northwest Branch National Veterans Home, located just west of Milwaukee and east of what was to become West Allis, may have been the cause of my father's family's settling in the Milwaukee area, but its influence went well beyond that—as we shall see.

At her 1900 marriage, Catherine was twenty-eight, and John Samuel Spearman, a sometime carpenter, was *fifty-one*—a Canadian-born, New York Civil War veteran widower, only ten years younger than Catherine's father; and who, like her father, had also been *living at the Veterans Home.*

In August, 1901, Catherine's widower father, William Lark, died at the Veterans Home in Milwaukee. During the Civil War, he had been in an elite unit, outfitted with Sharps rifles, called the *Illinois Sturgis Rifles*, which acted as Union Commanding General George McClelland's bodyguard. William Lark participated in the Peninsula Campaign and got ill at Savage's Station, Virginia in 1862. He had been a member of the G.A.R (Grand Army of the Republic), which, on his death, gave his daughter Catherine a letter saying his "genial smile" would be missed and his conduct as a husband, father, and comrade was "beyond reproach."

Two months earlier, about a year after their marriage, Catherine and John Spearman had a son, George, born where they were living, just south of the Veterans Home property, in what today is West Milwaukee.

Then the family moved to *Hot Springs, Arkansas*—for reasons I have never discovered, possibly related to John Spearman's health, or for work as a carpenter. Hot Springs, located in the Ouachita Mountains, south of the Ozarks, was famous for its waters, both for drinking and bathing, and was also known for its gambling casinos, racetrack, spas, bathhouses, bordellos, and many saloons. The Pittsburgh Pirates did spring training there as a team, but many baseball stars at the time came for preseason training. It was home to the Army and Navy Hospital since 1887, where active-duty military and veterans could come to enjoy the curative waters. Undoubtedly, it was here that John Spearman—whose Civil War unit was Company L of the New York Volunteer *Cavalry*—

introduced his stepson Billy Boulton to horse racing, which became a lifelong passion.

Anyway, it was in wicked Hot Springs in 1904 that Dad's mother, my grandmother, Rosaline Spearman, was born. A few months later, John Spearman was living at a Veterans Home in Bath, New York, not far from a son from his first marriage. Shortly afterwards, Catherine applied for half his veteran's pension on grounds of desertion.

A photograph of my grandmother and her brother was taken in Hot Springs in 1906, so Catherine was still there at that time, though no longer with her much-older husband.

That same year, 1906, back in Milwaukee County, Wisconsin, the four-year old village of **West Allis**, about five miles west of downtown Milwaukee, home of the state fair, was chartered as *a city*.

My grandmother Rosaline, age 2, and her brother George Spearman, age 5, 1906. (*Hot Springs, Arkansas*).

In 1900, Southside Milwaukee's Edward P. Allis Company, the country's leading producer of industrial steam engines and associated machinery used in lumber mills, flour mills, and mining operations, among others, had to turn down more orders than it could fill. The company, owned and run by the late founder's large family, had subcontracted three-year's work with a machine shop in Buffalo, New York and ran two satellite operations in Milwaukee called the South Foundry and the Bay State Works ("works" was a common term at the time for a heavy-industry factory). It's main 16-acre Milwaukee site on the Milwaukee Road rail line in the Walker's Point area near the port was officially called the "Reliance Works," but more commonly called the "Allis Works." To serve the two thousand, or so, employees of the factory—who worked a fifty-seven-hour, six-day, work week—the Milwaukee Road built a depot called the *Allis* station across the street from the three-city-block-long main Allis plant.

Wanting to enlarge and consolidate its operations, the Allis Company began looking at potential expansion sites in unincorporated areas of Milwaukee County. Land speculators, knowing the new factory would drive up local real estate prices, offered sites to the company at bargain prices. For $25,000, the company finally decided on a 100-acre site west of the city in unincorporated Wauwatosa Township, near where Catherine and John Spearman had been living—about halfway between the Veterans Home and the state fair park. The site was on the north side of *Greenfield Avenue,* where Dad was to grow up, which was the *town-line road* separating Wauwatosa Township to the north from Greenfield Township to the south.

The site was adjacent to the Milwaukee Road rail line which ran through southern Wauwatosa Township, and about a half-mile north of the Chicago & Northwestern line, which ran through the northern part of Greenfield Township. The land speculation deal that secured the site for the Allis Company involved a syndicate contributing cash and parts of three properties and included not only the site itself but also a guarantee of a railroad spur from the Chicago & Northwestern line, so the new factory would be served by *both* area railroads, a contingency that needed approval of both town boards.

The Allis Company was particularly attracted to the site because it was served by mass transportation. Two streetcar lines, originally intended for state fairgoers, would accommodate the company's Milwaukee employees—though the Allis Company hoped many of its employees would relocate near the new factory in residential neighborhoods soon to be platted in the anticipated local land boom.

In February of 1901, ground was broken for the "West Works" of the Edward P. Allis Company. To distinguish the development from the existing Allis Works, the new site was soon called the "*West* Allis Works." Real estate speculators, in the frenzy of buying and selling land around the factory under construction, began to call the whole vicinity "West Allis." Soon, other Milwaukee factories were announcing relocation to the new developing community called West Allis. The streetcar placards were changed to read "*West Allis.*" The Chicago and Northwestern railroad station, located around what is today 82nd and National Avenue, was changed from "North Greenfield" to "*West Allis.*" Likewise, the name of the North Greenfield Post office was changed to "*West Allis.*" In 1902, when the population of the area reached over a thousand, a group of citizens, mostly real estate speculators—with input from the Allis family—incorporated the community, including parts of southern Wauwatosa Township and northern Greenfield Township as the *Village of West Allis.*

Ironically, by that time, the Edward P. Allis Company no longer existed.

In May of 1901, while the anticipated two-year construction of the new factory was underway, the Allis company merged in a Wall Street deal with two manufacturing companies from Chicago and one from Scranton, Pennsylvania to form a conglomerate called the Allis-Chalmers Company (later reorganized as the Allis-Chalmers Manufacturing Company).

By the 1910 census, my great-grandmother Catherine Spearman had left Arkansas and was living in West Allis as a head of household with her four surviving children—two surnamed Boulton and two surnamed Spearman—calling herself John Spearman's *widow,* which she did the rest of her life. Meanwhile, my *very alive* great-grandfather John Spearman was living at the Washington State Soldiers Home in Orting, Washington. In the 1920 census, John Spearman resided on a farm in Potter County in Northern Pennsylvania, not far from Genesee, Pennsylvania, where he was buried when he died of heart disease in 1929 at eighty years of age—finally making Catherine the widow she had long claimed to be. Catherine applied to the newly organized Veterans Administration for a widow's pension after he died.

Of Catherine's four surviving children, two died before 1920. In 1916, her son George Spearman, my grandmother's brother, only fourteen, died of a kidney ailment called Bright's disease. *Critically for Dad,* in October of 1918, Catherine's oldest, twenty-six-year-old Marie "Mae" Boulton Schuster—mother of three daughters: Catherine, called "Kay," Margaret, and Gertrude—died in the *Spanish flu* pandemic. Mae's husband, John Schuster, died of the same flu fifteen months later—and my great-grandmother Catherine Spearman, at age forty-nine, became the *legal guardian* of her three orphaned granddaughters, the oldest of whom was seven. I assume Catherine and her three young wards lived primarily on general relief from the county. Catherine's surviving children were William Boulton and my grandmother, Rosaline Spearman, still in her teens.

In 1921, William Boulton, Dad's Uncle Bill (really, *half*-uncle)—the person who would have the greatest influence on Dad before the War—got married to Elizabeth "Betty" Rath. Applying for a marriage license, Bill, twenty-four, gave his occupation as "newspaper business." Betty, nineteen, was a telegraph operator. Until 1923, Bill and his new bride lived with his mother Catherine, half-sister Rosaline, and his three Schuster nieces at what was then 526 71st Avenue in West Allis, now 1418 79th Street. (West Allis changed its street numbering system in 1932).

By 1924, Bill was working as a men's clothing salesman at Espenhain's Department Store in downtown Milwaukee, and he and Betty moved to Milwaukee. Catherine, Rosaline—who was pregnant with Dad—and the three Schuster girls moved to 691-693 73rd Avenue in West Allis, today known as 1733-1735 81st Street.

It was there, in a storefront apartment attached to Roska's pharmacy at the corner of National Avenue, across the alley from Honrath's butcher shop, that unmarried, twenty-year old Rosaline Spearman gave birth to Dad the late morning of September 11, 1924.

I doubt Dad knew it, but where he was born had historical significance in his hometown. The building was constructed about 1896—before West Allis was founded—in a small business area on National Avenue, near the old North Greenfield station of the Chicago &

Northwestern Railroad, only a block away. Above the retail store (or stores) in the front of the building was a large, open room. When West Allis was first incorporated as a village in 1902, that second-floor room or hall was where the Village Board held its meetings—in effect, *West Allis' first Village Hall*. The building where Dad was born was also used as a post office for a time and was the location of the *first library* in West Allis.

So, how did Rosaline get pregnant? Again, the Veterans Home had a role.

When Dad was born, his mother was twenty and his father fifty. William Chandler Finney was admitted to the Veterans Home in Milwaukee on October 6, 1922, and he lived there for three years. On admission he was described as five feet eight inches tall with brown eyes and brown hair. He was Protestant and was a telegrapher. Dad's story was that his grandmother Catherine was in an auxiliary and used to volunteer at the Veterans Home where her father and husband had both once lived. I suspect she belonged to the Ladies of the Grand Army of the Republic, since her father was a G.A.R. member. At the Veterans home she supposedly befriended William Chandler Finney who was from a later war and five years her junior.

Dad hardly knew anything about his father, other than he was a Spanish American War veteran. The way Dad told it, William Chandler Finney was his grandmother's boyfriend who took advantage of his mother and then abandoned his mother and him. After so many years, the real facts of what happened and how it happened will never be known. But—thanks to the internet—I have learned many things Dad himself never knew.

Dad's father, my grandfather, William Chandler Finney, was the son of Dr. David Finney, a dentist in Logan, Hocking County, Ohio, located

Google Maps Street View of where Dad was born. (*©2017 Google*).

Front of the building where Dad was born, as it looked in the horse-and-buggy days. (*West Allis Historical Society*).[6]

near the Hocking Valley coal fields in the southeastern part of the state. Dr. Finney was a Civil War veteran originally from Elizabeth Township, Allegheny County, Pennsylvania, outside of Pittsburgh. He was active in Logan's Odd Fellow Lodge. When the Spanish American War broke out in 1898, his oldest son William—called "Will"—a twenty-four-year-old telegrapher and ticket agent for the Columbus, Hocking Valley, Toledo Railway, volunteered.

On July 5, 1898, William Chandler Finney enlisted in the Army at Columbus, Ohio, about fifty miles northwest of his hometown of Logan. The railroad for which he worked passed through Logan and Columbus on its way north to Toledo and Lake Erie. He was assigned as a private in the 13th Company of the Signal Corps of the U. S. Volunteers and was sent to the nation's capital. On September 4, 1898, he was at Washington Barracks (now Ft. McNair) and was sent by ship down the East Coast to Company 3 of the Signal Corps in Santiago de Cuba, a port on the southeastern end of the island. Most of the land and naval confrontations in the Cuban portion of the Spanish American War, including San Juan Hill, occurred around Santiago in June and July of 1898. The US and Spain signed an armistice ending hostilities on August 12. On September 18, 1898, when my grandfather's ship, after twelve days at

Google Maps Street View of the iconic 150-year old "Old Main" domiciliary at the Milwaukee Veterans Home. My grandfather, great-grandfather, and great-great grandfather may all have lived here—though some veterans also lived in barracks elsewhere on the property. (*©2019 Google*).

sea—and with a case of yellow fever aboard—anchored in the rainy harbor of Santiago de Cuba, the fighting was over.

My grandfather, William Chandler Finney, wrote an account of his arrival for the *Hocking Sentinel* newspaper—describing the disease-wracked city as a "tomb of living skeletons" and "filthy" with a "sickening odor." His unit, much depleted by disease, living in tents, was charged with stringing telegraph and telephone lines as part of the army of occupation. From October 23 on—only a little more than a month after his arrival in Cuba—he was sick. Yellow fever, malaria, typhoid, and dysentery were rampant in Santiago at the time. It should be noted, for every American killed in action in Cuba during the war, nearly ten died of disease. He was diagnosed with malaria, called "malarial poisoning" at the time. On November 6, he was put on a hospital ship to New York and given thirty days leave at home in Logan, Ohio. Having spent only five months in service, without ever returning to Cuba, he was discharged from active duty by telegraphic instructions to his unit in Santiago de Cuba on December 8, 1898.

In the 1900 census, twenty-six-year old William was living at the family home at 133 Zanesville Avenue in Logan, Ohio with his parents and his two older sisters, Estelle, a teacher, and Grace L., as well as his two younger brothers, Hayes W. and George M. Finney. Hayes, who had served in the Hospital Corps in the Spanish American War, was described as a dentist, but he seems to have made his living mostly as a male nurse, residing in Illinois, near Chicago. George was a telegraph operator like William, probably also working for the railroad.

Later that year, William was reported "dangerously sick with Cuban fever" in Springfield, Illinois, where he was working as a telegrapher. In 1902, my grandfather applied for a veteran's disability pension, based on his malaria. In 1904, he took a position as a train dispatcher at Chama, New Mexico. The following year, he was reported working as a Western Union telegraph operator and express agent at Indiana Harbor (East Chicago), Indiana. I think I found William in the 1910 census as William C. Finnie [sic], a railroad telegraph operator in Florence Township in Stephenson County, Illinois, near Freeport, west of Rockford.

On July 21, 1913, my grandfather, thirty-nine, married fifty-four-year-old Elizabeth Nickerson in Chicago, Illinois, with Rev. A.E. Ulrich of the Centenary Methodist Episcopal Church, 1230 W. Monroe Street, officiating. Bride and groom were both residents of Rolla, Missouri.

In September of 1918, my grandfather, who, according to an early newspaper account, played both the piano and violin, registered for the World War I draft. At that time, he gave his residence as Rural Route 2, Rolla, Phelps County, Missouri. He gave as his nearest relative his wife, Elizabeth N. Finney. His occupation was "Nothing."

I was not able to find him in the 1920 census, and I do not know how his marriage ended, by death or divorce; but when he came to the Veterans Home in Milwaukee in 1922, he said he was single. His $8 per month pension was shifted from the address in Rolla, MO, so he apparently lived there up to the time he came to Milwaukee.

In 1921, William's father, old Doc Finney, died in Logan, Ohio at age seventy-eight of rheumatic arthritis. The widow, my great-grandmother, Augusta Sarah (Murphy) Finney, and daughter Grace moved from Logan to Columbus, Ohio, where the youngest son George was a trainmaster, putting together freight trains for the C&O Railroad, which had bought out the old Hocking Valley rail line. The Finneys bought a two-story five-bedroom house at 701 Linwood Avenue in Columbus that stayed in the family after the widow Finney's death

The only photo I have of my grandfather William Chandler Finney, left, taken with his younger brother Hayes W. Finney in Chicago, August 1926.[c]

in 1929, and until the 1950s. This address was where two or more of William's siblings lived at any given time, and where William himself, last survivor of his family, was living when he fell at home and broke his hip at age eighty-two in 1956 and died of kidney failure two months later. I have found no evidence any of his four siblings ever married or had children.

Google Maps Street View of 133 Zanesville Ave., Logan, Ohio. (©Google 2017).

The parents and all five Finney children, including my grandfather, are buried at Oak Grove Cemetery in Logan, Ohio. My grandfather was married twice, but his death certificate said he was "never married." At the time of his death he had a son, Dad, and three grandchildren, but a brief obituary in the Logan *Daily News* said he had "no immediate survivors." Dad, of course, had no memory of his father and was given no notice of his father's death.

As for Dad, on September 18, 1924, only one week old, he was baptized at Holy Cross Catholic Church in Milwaukee, located on Bluemound Road, near the Veterans Home—not at Holy Assumption church in West Allis where his grandmother and mother attended and where his mother had gone to grade school. Dad probably never knew it, but, according to the church's register, he was baptized with his mother's surname as William Norbert *Spearman*. (Dad seemed unsure as to how he got the middle name Norbert but thought it possibly had something to do with the Norbertine priests). His godmother and only sponsor was his grandmother, Catherine.

I have a certificate from ten years later, dated March 26, 1934, attesting to Dad's baptism as the child of William Finney and Rosaline Spearman and stating Dad's name as William Norbert *Finney*. The 1934 certificate I have, apparently obtained by his Uncle Bill, was issued so

Dad could make his First Communion, which he did on May 13, 1934, at Holy Assumption Church in West Allis, where he was attending parochial school at the time.

I do not know if my grandfather attended the baptism, but he may have, since Holy Cross was walking distance from the Veteran's home. I doubt Dad's Uncle Bill was at the baptism. I have a letter from William Chandler Finney to Uncle Bill from 1926 that suggests Uncle Bill was angry to learn of the baptism and the naming. Dad believed he was named for his father, though his uncle, of course, was William, as well.

A bit after the fact, but on December 9, 1924—about three months after Dad was born—William Chandler Finney and Rosaline E. Spearman were *married* in a civil ceremony in Milwaukee, in front of a judge, probably at the court house. This, too, seems to have been done behind the back of Uncle Bill. In the same letter, my grandfather accused Uncle Bill of trying to deny the existence of the marriage despite the legal formalities. Uncle Bill supposedly referred to his half-sister as Rosaline *Spearman,* though she went by the name Rosaline *Finney* the rest of her life.

On February 13, 1925, approximately *five months after the birth,* a birth certificate for Dad—with a last name of Finney—was recorded at the West Allis Health Department. In the box marked "Legitimate?" was put a "Yes." The residence of the parents was given simply as "West Allis, Wis." The father's occupation was given as "clerk." Dad's birth had been attended by J.J. Wilkinson, M.D.

I do not think Dad was aware his father lived in Milwaukee at the Veterans Home for over a year after Dad was born. What occurred during that time is hard to say. Uncle Bill and wife Betty lived in Milwaukee, where both worked downtown. Betty became a telephone operator for the prestigious Pfister Hotel. Bill went from salesman to department manager at Espenhain's Department Store. Bill was always a fastidious dresser and worked in men's clothing, making occasional buying trips for the store to the garment district in New York. Although Bill no longer lived with his mother, he almost certainly contributed financially to the support of his mother, his eight-year-younger half-sister and infant, and three young nieces—all on welfare. He did not like his new brother-in-law, a veteran—like the stepdad who abandoned his mother and him—and old enough to be his father. The old soldier, William Chandler Finney, kept on living at the Veterans Home on his pension.

• • •

Dad was born in the Roaring Twenties during Prohibition. Women had been given the vote, and many bobbed their hair, raised their hemlines, put on silk stockings and experimented with smoking, drinking bootleg liquor, and attending wild dances with jazz orchestras.

Dad was never specific, but he told me Uncle Bill had *underworld* (i.e. gangster) connections. Prohibition, of course, was never popular in Milwaukee with its heavy foreign population and its beer-loving reputation as the most German city in America. In light of Bill's later occupation as a bartender, it is possible he worked part-time in a speakeasy. Or he may have met bootleggers while selling them the

Two of Dad's Schuster cousins, Gertrude and Kay.[c]

latest in men's clothing at the department store. However Uncle Bill's underworld contacts were formed, he apparently used them to pressure my grandfather to stay away from Rosaline.

In 1926, Dad's father William Chandler Finney sent a letter to Uncle Bill—with a copy to Rosaline, which she kept—that made an apparent reference to Dad's orphan cousins, the Schuster girls.

A bizarre quote from the letter to Uncle Bill gives a flavor of the situation:

> *"... Do I have to get your permission in this 'Glorious Land of the Free' to upbraid my wife for her alleged indiscretions and association with other men, and those members of an alleged gang of crooks and criminals; some of whom threatened me with death if I did not keep my_mouth shut? ... Is your alleged hatred for American soldiers so strong, allegedly, that you approve of the alleged exploitation under cover of the blackness of night of a deceased veteran's orphan children [grandchildren? Dad's cousins?] in one or more Road House Dance Halls for the alleged entertainment of allegedly bleary eyed midnight revelers and rowdies? Do you derive anything from the supposed profit; or does it all go to the alleged scheming old woman and her son and his wife who allegedly ran these dances of the doubtful sort; allegedly taking tired sleepy children in an automobile at the Devil's hour and whizzing them to associate with Negro jazz undesirables? The children dressed like cabaret freaks. If you can bar me from my legal child [Dad] is he to be mistreated allegedly likewise? ..."*

Whether or not he was truly in fear of his life, William Chandler Finney left the Veterans Home in Milwaukee in November of 1925, probably to live with his brother Hayes in the suburban Chicago area. The letter he wrote to Uncle Bill in June of 1926 was from *General Delivery, Chicago* and complained of his not having heard from Rosaline due to Bill's influence. So, by that time, Dad's father was already out of his life. As of 1930, my grandfather was in a Veteran's Home in Johnson City, Tennessee. I have found no evidence any pension application was ever filed for Dad based on his father's military service. While growing up, Dad received no financial support from his father, and never heard from him. Dad always spoke of his father with a sense of *bitterness* at what he viewed as abandonment.

Unfortunately, something even worse than being abandoned by his father was to happen.

Dad's mother, Rosaline Spearman Finney, was a victim of *grand mal* epileptic seizures. I do not know if she was epileptic from birth, though that seems to have been the case. As a young boy, Dad was trained how to act in the event his mother had an episode. He was to try to prevent her from injury while she became unconscious and her body went into violent spasms for several minutes. He said afterwards, his mother, having no memory of the attack, would ask him, "Did I have one of my spells?"

When Rosaline was a very young girl, no medication was available to prevent or control epileptic seizures. Children having the affliction often were also subject to anxiety and depression. Epilepsy brought with it embarrassment and social stigma since the other children, and most adults, did not understand it and experienced fear and helplessness when confronted by a child having an attack. The barbiturate Phenobarbital, invented in Germany, came to market in 1912 but was unreliable and caused sleepiness and behavioral issues.

Rosaline attended Holy Assumption Grade School in West Allis through 8th grade. I believe in those days the public schools had junior high schools (grades 7–9) and senior high schools (grades 10–12). The 1940 census asked the highest level of education completed, and for Rosaline it was one year of high school, so probably 10th grade. Dad said his mother attended West Allis High School.

The 1923 West Allis City Directory listed her occupation as "packer," whatever that means. I have found no other evidence of her working. I suspect her epileptic condition limited her employment opportunities. She most likely helped her mother with the three little Schuster girls, Rosaline's nieces.

By 1927, with Dad still an infant, his grandmother Catherine, her daughter Rosaline, and the four grandchildren moved to an address that would today be 7132 W. Greenfield Avenue in downtown West Allis, almost certainly above a store, either a paint store or a small grocery store.

The huge—by this time 160-acre—*Allis-Chalmers West Allis Works* on W. Greenfield Avenue ended at 70th Street. It employed some 15,000 factory workers and manufactured everything from farm tractors to huge steam and hydro turbines and electrical generators, transformers, mining and factory equipment, etc. The nearly 200-acre *Wisconsin State Fair* Park started at 77th Street. The seven blocks of W. Greenfield Avenue in-between had become a bustling commercial center, the main street of downtown West Allis.

Dad lived in or near there until he went into the Army.

He told me of one incident of neglect while he was a toddler. He was left outside in the hot sun on a second-story roofless back porch and suffered sunstroke. Fortunately, an icewater bath was able to bring his temperature back down.

From what I can determine, in 1927 Rosaline's behavior became strange and frightening. According to her death certificate, from many years later, Rosaline suffered from "epileptic *psychosis*" dating from April 4, 1927. On that date, some incident most likely occurred that prompted her mother to start legal proceedings to have Rosaline declared insane. I suspect Catherine's concern was Dad, only about two and a half at the time.

The upshot of it was that Dad's mother, twenty-three-year old Rosaline, was committed to the Milwaukee County Asylum for the Chronic Insane in Wauwatosa, where, with the exception of a few home leaves, she spent the rest of her life, dying during an epileptic attack at age forty-eight in 1952.

I think her funeral was on my fifth birthday. I vaguely remember visiting her a few times at the Milwaukee County Institutions as a child with my parents and sister Mary. We called Rosaline "Grandma-in-the-Hospital."

I wish I could remember her better. My mother told me Rosaline was an intelligent woman who read the newspapers and kept up on current events. I remember the asylum as a spooky place, but Rosaline herself as okay, though she wore filmy dresses, like from the 1920s and 30s, not like the fashionable 1950's clothes Mom wore.

Dad told me several times Rosaline probably would have been let out of the asylum except there was no one to take her out after her mother died. On at least one occasion, I assume while Catherine was still alive, Rosaline had a home visit and apparently had a tantrum. Dad recalled her screaming and throwing things. Especially angry with God and the Catholic Church, she took a crucifix off the wall and threw it.

She was quite upset when our family moved to Green Bay in 1951. Green Bay was only a two-and-a-half hour drive from Milwaukee at the

Rosaline, age 13, 8th Grade graduation at Holy Assumption School.[c]

time, but to her mind, still stuck in the 1920s, it might as well have been at the end of the earth.

With Rosaline institutionalized, Dad's grandmother Catherine, at fifty-seven, became his legal guardian, as she was for Dad's three cousins, the Schuster girls, the oldest of whom, Kay, just entering her teens—and all of them apparently living on county relief.

In 1929, the Depression started taking hold. Uncle Bill lost his job at Espenhain's, which had gone bankrupt after fifty years in business. He was described in the Milwaukee City Directory as a salesman at Gimbel's Department Store. In the 1930 Milwaukee Directory he was listed simply as a salesman, and he and his wife Betty had separate residences.

By 1931, Bill had moved back with his mother—who had cancer—and his three nieces and nephew on Greenfield Avenue in West Allis. I do not know why childless Bill and Betty got divorced in April of 1931—though Bill had an eye for the ladies. Dad apparently liked Aunt Betty and I think kept in touch with her over the years.

Dad and his mother, Rosaline.

Latest photo Dad had of his mother, my grandmother, Rosaline Finney, age 25, Ward 4, August, 1929.[c]

2

THE DEPRESSION

Dad went to kindergarten at public school and then started grade school at nearby Holy Assumption in West Allis, from which his mother had graduated. The parish was founded in 1902, before West Allis even became a city. Its young, dynamic pastor, Reverend Julius H. Burbach, was a driving force in the community and became the city's first historian. (In the parish's 50th anniversary memorial book, Dad was listed as one of 465 former Holy Assumption students who served in uniform during World War II—14 of whom were killed in action).

Dad told me an anecdote about his school days at Holy Assumption that I am sure happened to others. The School Sisters of Notre Dame, a German order, were not above corporal punishment to keep the kids in line. On one occasion, Dad wrote an essay that contained some words not to his teacher's liking. The nun made Dad stand beside his desk, and she walked up to him with a metal-edged ruler in her right hand. "Show me the hand that wrote that composition," she said. Dad timidly extended his right hand, and the nun took it, palm up, in her left, raised the ruler in her right and brought it down forcefully. But Dad told me before the metal edge could hit its intended target he pulled his hand away, and the ruler came down *on the nun's own hand*. Dad proudly said he "really caught hell" after that—but he was always *glad* he did it.

Dad was a precocious child, obviously bright, and liked to sing. He said Al Jolson was his favorite singer. Dad used to sing—I assume for money—as a very young child on street corners in downtown West Allis, sometimes with a friend named Billy Wong, who later became a dentist. If I have the story right, Dad sang once in a talent competition at the Riverside Theater in downtown Milwaukee; and a talent agent wanted to sign him up and take him to Hollywood to become a child star, but Catherine, or Uncle Bill, would not allow it.

The year 1931 is usually considered the worst year of the Depression. It brought an especially unfortunate turn of fate in Dad's young life. His grandmother and legal guardian, Catherine E. Lark Boulton Spearman, developed breast cancer and died in December at age 61. She left a note to Uncle Bill with the deed to the family cemetery plot in Evanston, Illinois where her grandparents and three deceased children were buried—adjacent to the cemetery plot owned by her surviving brother George Lark, where her Civil War veteran father and the mother she lost when only thirteen were buried. She ended the note with these words: "Watch over your sister's [Mae's] children and Billy [Dad] too, as that is what's killing me, to leave you all. You don't know what I've suffered. Mother. *Pray for me.*"

I remember Dad telling me his grandmother used to wear green stockings on St. Patrick's Day and would dance the Irish jig. Only seven when she died, he always spoke of her—the nearest thing he had to a mother—with great affection.

Instead of letting his nephew and nieces go to an orphanage, Bill Boulton, at thirty-five, followed his mother's wishes and took on the role of legal guardian for Dad and Dad's three female cousins. The county helped, too. I remember Dad telling me, during the Depression, he could tell when other kids were on relief, like he was, because they also wore corduroy knicker pants, like he did. Dad seldom talked of it and always seemed somewhat ashamed of having grown up on county relief. However, I suspect Uncle Bill also contributed significantly to Dad's support over the years.

Uncle Bill was a bit of a mystery man. What he actually did for a living during the Depression I do not know, but he worked at least part-time and probably full-time as a *bartender*. While Prohibition did not end until 1933, Wisconsin was one of the most rebellious states when it came to enforcement. A 1931 federal report called it "a Gibraltar of the wets" and a "wet Commonwealth."

Despite the efforts of federal agents, the success of Prohibition depended on *local law enforcement*. Although federal law allowed no alcoholic beverage of over .5 percent alcohol, Wisconsin passed a law in 1926 that made legal the manufacture of beer with an alcoholic content of 2.75 percent. The above-mentioned federal report said of the over 1,200 "soft-drink parlors" in Milwaukee's telephone book, *all* sold liquor (i.e., while local authorities looked the other way).

Wisconsin overwhelmingly passed a referendum in April of 1929—four years before the end of Prohibition—that led to repeal of the state's

Dad on Greenfield Avenue about 1929 (age 5).[c]

Uncle Bill Boulton tending bar.

"dry" enforcement statute, the Severson Act. That meant local law enforcement did not have to enforce Prohibition at all. So, Bill could have worked legally as a bartender for years—at least under state law—even before Prohibition was officially repealed under federal law in December of 1933, by the *21st Amendment* (authored, incidentally, by Wisconsin's U.S. Senator John J. Blaine).

In 1934, after Prohibition ended, the American Legion built the Tanner-Paull Post, No.120, at 6922 W. Orchard Street in West Allis, a block south of Greenfield Avenue, and that is where Bill definitely worked as a bartender most of the time Dad was growing up. As a bartender in those days, Bill was like an entertainer, good at retelling the latest jokes and funny stories—a skill he passed on to Dad.

Although the West Allis City Directories in the 1930s did not list Bill, I have no reason to suspect he did not live in West Allis around West Greenfield Avenue. He seemed never at a loss for female companionship and likely stayed often with women he knew. Meanwhile, Dad was attending Holy Assumption Grade School, located at 72nd and Orchard, two blocks from the Tanner-Paull post where Bill worked, and a block south of Greenfield Avenue.

West Allis, fifteen minutes away from downtown Milwaukee by interurban line, was perhaps most famous as the home of the Wisconsin State Fair Park, on West Greenfield Avenue, between 77th and 84th

Streets, permanent site of the State Fair since 1892. State Fair Park was also the home of the *Milwaukee Mile*, an originally dirt (later paved) auto race track, the Milwaukee area's premier car-racing venue since 1903—and reportedly the oldest continuously-operated motor speedway in the world. Taking advantage of the 37,000-seat grandstand at the race track, the Green Bay Packers football team from 1934, and all during the 1940s, played some of their home games at a football field laid out in the track's infield. The eyes of football fans everywhere were on West Allis when the NFL Championship game was played there in 1939 between the Packers and the New York Giants. (Packers won 27–0).

As mentioned, two railroads served industrial West Allis. With factories such as Allis-Chalmers, Kearney and Trecker, Pressed Steel Tank, West Allis Malleable Iron & Chain Belt Company, and numerous others, West Allis had become Milwaukee's biggest blue-collar suburb; many of Dad's friends had fathers who worked—when work was available during the Depression—"in the shop."

The Roosevelt administration helped Depression-era West Allis. On West Greenfield Avenue, near City Hall, the Works Progress Administration (WPA) built the main post office (featuring murals by Chicago artist Frances Foy). Another WPA project was construction of a channel lined with "Lannon stone," quarried in neighboring Waukesha County, to control flood-prone Honey Creek. Allis-Chalmers won Navy contracts and also built some of the monstrous hydroelectric turbines for the federal Boulder Dam project near Las Vegas—known today as the *Hoover Dam*.

As the main thoroughfare of West Allis, West Greenfield Avenue afforded public transportation. Streetcar tracks ran down the middle of it. It had two movie theaters, the *Allis* and the *Capital*, and a third theater, the *Paradise* opened at 63rd Street. Greenfield Park was developed near the county line (124th St.). The park had a swimming pool (built by the WPA) and a golf course. I don't know where Dad learned to swim—possibly at one of the many lakes around the Milwaukee area, at Greenfield Park, or at West Allis High School's indoor pool. He visited Milwaukee's Lake Michigan beaches in summer.

He also learned to golf, probably at Greenfield Park. During summer vacation, he made money as a caddy at the Bluemound Country Club.

In the summer, Dad visited Washington Park in Milwaukee, which

Dad in corduroy knicker pants with friends on Greenfield Avenue.[c]

One of my favorite pictures of Dad as a teen.[c]

had a zoo and the Blatz bandshell. Dad also attended baseball games at Milwaukee's Borchert Field, home of the Milwaukee *Brewers*, a minor league team owned by the Boston *Braves*.

After completing 8th Grade at Holy Assumption, Dad, a bright but not disciplined student, switched to public school and attended John Dewey Junior High School for 9th Grade before going to West Allis High School at 75th and National Avenue. Slight of build, he was quite agile and would amuse his friends by doing acrobatic back flips.

Dad told me he liked radio comedies like "Fibber McGee and Molly,"

West Allis (later West Allis Central) High School, opened in 1921. (*West Allis Historical Society*).[c]

Dad (far left) with friends John Hammer, Jack Jung, Don Andrews, Rich Bernier and Paul Franson, core of "The Gang"—all of whom ended up in service in World War II.

dramas like "The Shadow" and "Gangbusters"; and he was a fan of "The Green Hornet" both on radio and as a movie serial. He also liked "Our Gang" comedies at the movies.

During the Depression, Dad made occasional trips to Chicago with Uncle Bill to visit his great-uncle George Lark, his late grandmother's brother, and to attend the horse races with Uncle Bill at Arlington Park. Dad told me he was only seven when he smoked corn-silk cigarettes and soon graduated to the real thing, smoking for the rest of his life. He got sick on home brew as a kid; and—unusual for someone from the Milwaukee area—he never drank beer again.

While Dad's family life was not traditional, and likely would be called *dysfunctional*, he managed to make life-long friends in West Allis, who eventually called themselves "The Gang." The parents of Dad's friends all sympathized with Dad's orphan-like circumstances during the Depression and went out of their way to be kind to him.

For example, Dad often spoke fondly of the parents of his friend Arnie Friedman who frequently had Dad over for dinner and gave him a life-long appreciation for cream soda.

Uncle Bill always seemed to have a car, which gave Dad certain mobility. Dad apparently learned to drive—at age 14 he once told me—but I don't

Uncle Bill with keys to what appears to be a 1939 Dodge.c

think he got a license until after the War. At least I remember him telling me he had to take a driving test when he got out of the service, and it consisted of getting the tires stuck in the streetcar tracks on Greenfield Avenue—and then getting out. (Road tests were required by West Allis under *municipal option* at the time. The state itself did not actually require road tests until 1956).

Dad took *Drafting* in junior high school, a popular subject no doubt in blue-collar West Allis. He also had to take "shop" courses in wood and metalworking. He said the most useful course he took in high school actually was *Electricity*. I know he was not afraid to do wiring and install outlets and switches related to the many do-it-yourself projects he did in the homes where we lived. Dad was also very good at math. He could figure in his head what most people had to write down.

According to the 1940 census, West Allis, with a population of over 36,000, was the largest suburb of Milwaukee, the country's thirteenth largest city. Dad, at that time, still in high school, was living at 7209 West Madison Street in West Allis, two blocks north of Greenfield Avenue, with Kay and Ray Blazejewski—the surname was later changed to *Blake*. Catherine, or *Kay*, named after her late grandmother, was the oldest of the three orphaned Schuster girls who were Dad's cousins and grew up with him. She was eleven years older than Dad, and later became my godmother. Her younger sisters, Margaret, who was somewhat emotionally challenged, and Gertrude, both in their twenties, had by then moved away.

Dad was listed as a "nephew" in the 1940 census, but Uncle Bill, who

Dad, mugging in a photo booth.[c]

Dad and his cousin Kay.[c]

officially lived with Dad and Kay and Ray, was not listed. I suspect that was because he was not in West Allis on the official enumeration day, which was April 1. Dad told me, for a number of years, Uncle Bill used to drive to vice-mecca Hot Springs, Arkansas, where he once lived as a boy, and work at the horse-race track there during the racing season. The Oaklawn track traditionally was open for racing from mid-January to mid-April. Photographs from 1938 and 1939 show Uncle Bill during those years visited Hot Springs—called by *Liberty Magazine* in 1939 "the wickedest city in the United States"—and I suspect he did in 1940, as well.

With a birthday in the first half of September, Dad was always one of the youngest in his class at school, and one of the smallest. He had to deal with bullies and taught me a thing or two from his experiences. "Bullies are basically cowards," he used to tell me. "They pick on the small and the weak because bullies themselves are actually afraid of getting hurt. What you have to do is stand up to them and fight them. If you don't, the bullying will never stop. Don't be afraid to fight them. Just make sure you *hurt* them, even if you get beat up. As long as they know you can hurt them, they will leave you alone and pick on someone who won't fight back." That is what Dad told me he learned on the streets of West Allis during the Depression.

One time Dad started fighting a bully, and the kid ran into a store, thinking he would be safe. Dad wanted to show he could hurt the kid and did not let up. He followed him into the store and kept fighting

Gag photo Uncle Bill sent from Hot Springs, Arkansas in 1938 as a "bar man."

until people in the store broke up the fight. Dad said the kid never picked on him again.

During the Depression, in West Allis, as elsewhere, the grocery business was undergoing a revolution. Small walk-in stores, with all the groceries behind counters, were giving way to larger so-called "self-serve" chain stores, with cheaper prices. The automobile was changing the way people shopped. When Dad was in junior high, a New York chain, the Great Atlantic and Pacific Tea Company, more familiarly known as *A&P*, the country's largest grocery retailer, closed all but two of its seven small stores in West Allis, and in September of 1938 opened a larger, one-stop store it called a self-serve "super-market," across from City Hall, at 7524 W. Greenfield Avenue, a few blocks from where Dad lived.

Besides the usual grocery items, the new store sold fruit and vegetables, as well as baked goods, and took advantage of advances in refrigeration to sell dairy, meat, and poultry. A&P was a pioneer in the development of less expensive "house brands" such as *Ann Page* jarred, canned, and boxed products, *Iona* canned vegetables, *Jane Parker* bread and baked goods, *Our Own* tea, and fresh-ground *Eight O' Clock*, *Red Circle*, and *Bokar* coffee, whose aroma permeated the store. The shopper could push the recently-invented shopping cart—A&P initially called their basket-

This high-contrast copy of a photo from 1965 is the only image I could find of the A&P supermarket on Greenfield Avenue where Dad worked before the War. (Courtesy West Allis Historical Society).

on-wheels a "glider"—through the wide aisles, selecting what to buy, pick up a *Woman's Day* magazine, sold exclusively at A&P, at the check-out stand—and take the new purchases to the family car in the adjacent angled parking lot.

A&P hired Dad part-time as a clerk for the Greenfield Avenue store while he was still attending West Allis High School, only three blocks away. Among the skills Dad learned at A&P was how to operate a cash register and an adding machine, what we today would call 10-key.

While Dad was in high school and working at A&P, always in a starched white shirt with black tie—usually a clip-on bow tie—he was aware of the wars going on in China, Europe, and North Africa. In fact, wars seemed to be going on everywhere.

The Versailles Treaty after World War I, patterned in part on the *Fourteen Points* of U.S. President Woodrow Wilson, had set the stage for World War II.

During the 1930s, Germany's dictator, Adolf Hitler, incrementally violated many of the Versailles Treaty's terms. In defiance of the treaty, he rebuilt and modernized the German military. In 1936, his army occupied and fortified a treaty-created "demilitarized" border zone with France, Belgium, and the Netherlands called the Rhineland—and the Allies, facing domestic issues in the world-wide Depression, did nothing. In March of 1938, he invaded Austria in a more or less bloodless takeover done in the name of unifying the German people, completely contrary to the treaty—and the Allies did nothing. In September of that year, in violation of the treaty, he threatened part of Czechoslovakia called the Sudetenland, which contained a lot of German-speakers from the old

Austro-Hungarian Empire—again in the name of unifying the German people. This also was a more or less bloodless takeover after he promised the Allies Germany had no more aspirations for land in Europe. So the Allies again did nothing. In March of 1939, convinced the Allies would avoid war at all cost, despite his previous promises, and again in violation of the treaty, he took over the rest of Czechoslovakia. The Allies still did nothing.

The Treaty of Versailles provided for the restoration of a Polish homeland with a guarantee of access for Poland to the Baltic Sea. That access, the so-called "Polish Corridor," was a strip of land that physically separated Germany's state of East Prussia from the rest of the country, much in the way West Berlin was cut off from West Germany during the Cold War. More fearful of the Soviets than of the West, Hitler made a nonaggression pact with the USSR, offering part of Poland if the Soviets would approve of an invasion to reconnect East Prussia to Germany. Hitler had hoped the British and French policy of "appeasement," or *giving-in* to avoid war, would continue.

But the German invasion of Poland on September 1, 1939 is officially what started World War II. Britain and France finally declared war on Germany, *but not on the Soviets*, who invaded eastern Poland—as well as the Baltic countries—a few weeks later.

. . .

On May 10, 1940, after conquering Denmark and Norway (critical to transport of neutral Sweden's iron ore), the Germans launched their *Blitzkrieg* attack on the Low Countries and France. The same day, Winston Churchill became Prime Minister of Britain; and Britain, to secure its interests in the North Atlantic, attacked—in as friendly a way as possible—strategic Iceland, which wanted neutrality and had refused to join the Allies.

From the time when they declared war on Germany in September of 1939 to the time of the May 10, 1940 attack by Germany, France and Britain had done only sporadic fighting against the Nazis. The period is usually referred to as the "phony war." (Hitler actually wanted to attack

France right after the defeat of Poland—but believed his army was not suited to winter fighting).

France and Britain used the time to prepare for a defensive battle, anticipating a *counterclockwise sweep* of German forces through Belgium, as had occurred in World War I. Under the established tradition that an attacking army had to be at least three times larger than a defending army, the Allies, numerically superior in infantry, armor, and aircraft, were confident any attack by the Nazis would be repelled.

Instead, the Allies—making the classic mistake of preparing for the *last* war instead of the *next* one—faced a breathtaking onslaught very different from World War I.

Unlike the German command structure, which emphasized opportunistic decision-making by the officers closest to the action, the highly-centralized French command structure, to which the British Expeditionary Force was also subject, left little discretion to local commanders—who had to request authority from Paris and hope for a timely response from a superior far from the action. The French and British believed armor should basically support infantry and had *no separate armor divisions* to counter the spearhead German Panzer divisions. Much of the French and British aircraft were destroyed on the ground in the early days of the attack. Allied air power had been mostly designed to support *infantry*. The Germans, on the other hand, used their accurate Stuka dive bombers ahead of their *armor*, and the *Luftwaffe* was used not only tactically, but also to destroy the ability of the Allies to reinforce and supply.

The French high command largely ignored intelligence that disclosed the Germans were moving their main force through the hilly Ardennes Forest. The French believed activity in the Ardennes was only a diversion. The German attack in the Low Countries was successful, but was only a prelude to the armor, mobile artillery, and mechanized and motorized infantry main attack through the Ardennes, which caught the French and British completely by surprise. The German *clockwise* sweep, heading for the North Sea, left the French and British troops encircled.

We know today, the speed with which the attack took place was in part due to the German use of a drug called *Pervitin*, a methamphetamine that enabled attacking German soldiers to go days without sleeping. The

French had also foolishly committed their reserve troops in Belgium and had *no strategic reserve* on which to draw for a counterattack.

In less than a month, most of the fighting in France was over. The "miracle of Dunkirk" evacuation from France to Britain, ending June 4, saved nearly 350,000 British and French soldiers—who left most of their equipment behind. By June 25, nearly 200,000 more British and Allied troops were rescued from western and northern French ports by the British Navy with RAF protection.

On June 10, 1940, as the French government, facing surrender, was abandoning Paris, Italy *declared war* on France and Britain and soon invaded French territory. President Roosevelt was clearly alarmed at Italian dictator Benito Mussolini's choice of *force over diplomacy* in siding with the conquering Nazis. Roosevelt had tried to keep Italy out of the war and neutral—like Fascist Spain and Portugal. Speaking at commencement exercises at the University of Virginia, Roosevelt said he had warned Mussolini war in the Mediterranean would prejudice the "ways of life and government" and "the trade and commerce of *all* the American republics." In response to Italy's entry into the War, FDR then announced pursuit of two simultaneous courses of action. First, he said, the material resources of the US would be extended "to the *opponents* of force" (i.e., to Britain and France). And second, at the same time, he vowed, we would develop *our* resources and equip and train *ourselves* for "any emergency and every defense." (Within days, controversial legislation for the first peacetime draft in US history was introduced. After three months of debate, it passed and was signed by the president).

By the end of June, 1940, the British army had been driven from the continent; and the Nazis had conquered Denmark, Norway, Luxembourg, Belgium, Holland, and France—part occupied, and the rest governed as a puppet state from Vichy as of July 10.

Mussolini became greedy. He anticipated the Germans would defeat the British as quickly as the French, and he wanted Italy to have a stake in any resulting peace talks that might involve the Mediterranean and Africa. He hoped the Germans would reward Italy for tying down British troops outside of the British Isles.

In early August of 1940, from its colony Italian East Africa, and from the Vichy colony of French Somaliland (today's Djibouti), Italy attacked

British Somaliland in the Horn of Africa. In September, Mussolini's troops struck in North Africa, attacking Egypt from the neighboring Italian colony of Libya. In October, Italy attacked Greece in the Mediterranean. England, fighting for its life in the air war against the *Luftwaffe* known as the *Battle of Britain*, or the *Blitz*, was bound by treaty to defend both Egypt and Greece.

After the rapid fall of France, US newspapers and newsreels shown in all the movie theaters told the story of the valiant English standing alone against the Nazis and Fascists, who now controlled much of the European continent (though Fascist Spain and Portugal remained neutral). Films, photographs, and newspaper and radio accounts about London during the *Blitz* made most Americans admire the courage and determination of the British people, who, despite their defeat on continental Europe, were fighting the Italians and Germans in the Mediterranean and in North and East Africa—while getting bombed by the *Luftwaffe* at home.

By 1940, the 1937 Japanese invasion of China was old news, as was the 1939 German-Russian invasion of Poland, which had caused Britain and France to declare war on Germany. Milwaukee had a large Polish community on its South Side and was keenly interested in news about the "old country."

The fall of France was a wake-up call for the United States, always concerned about Japan in the Pacific and wary of how much the Vichy French government, with territories in the Caribbean, would collaborate with Germany. In July of 1940, Congress passed a massive spending bill popularly called "The Two-Ocean Navy Act." In early September, President Roosevelt, who had been assistant secretary of the Navy during World War I, signed a deal whereby the US—though still officially neutral—would transfer to the Brits fifty old, but desperately-needed destroyers (mostly for convoy duty) in exchange for leases to build naval and air bases on British territories in the Caribbean. For good measure, Winston Churchill, in the *Destroyers for Bases* agreement, even included Bermuda, in the Western Atlantic, and Newfoundland, in Canada, as sites for US bases, to protect future convoys from North America.

The same month, Italy and Japan—enemies of Germany in World War I—signed a mutual defense treaty with the Nazis called the *Tripartite Pact* (later joined by Hungary, Romania, Bulgaria, Slovakia, and Croatia).

In November of 1940, the British, in old-fashioned biplane *Swordfish* torpedo bombers, badly damaged much of the Italian fleet in the harbor of Taranto, in the arch of the Italian boot—the first time carrier-based aircraft successfully attacked a fleet in port. A team of naval planners was sent to Taranto to study the attack by Italy's new ally—Japan.

Several times a year, President Roosevelt made radio addresses to the nation called "fireside chats." On December 29, 1940, after he was elected to an unprecedented third term, Roosevelt in one of his "chats," fearful democracy itself was at risk, pushed for military aid to England and other countries fighting the Axis dictatorships—despite the position of the United States as a neutral country. He said he needed to warn the American people of the threat to their freedom posed by the Nazis, Fascists, and Japanese. If England were forced to surrender, and the Nazis took over the British navy, or the Japanese captured Hawaii, oceans would no longer protect us. Or what if the Nazis took over some other country or island in the Western Hemisphere [like Vichy French Martinique or Guadeloupe] as a base to attack us by air, as England was being bombed from France? He tried to prepare the American public for our participation in the War. Roosevelt called on the US to be the "arsenal of democracy" and emphasized the importance of the average factory worker making goods for the national defense—something well received by the people of blue-collar West Allis.

A few months later, Congress passed the Lend-Lease Act by which the US, though neutral, would furnish arms, ammunition, food, oil—whatever was needed—to any country fighting the Nazis, Fascists, and Japanese.

The president had to walk a fine line, however, since German-Americans, as in the Milwaukee area, were the country's largest ethnic group.

In spring of 1941, the big news in West Allis was the strike that had started January 22 at the Allis-Chalmers plant. The Milwaukee area's biggest factory, Allis-Chalmers—the headquarters and largest of its eight plants only two blocks away from where Dad lived—had a number of defense contracts, especially with the Navy. The leadership of the UAW-CIO local that went on strike in West Allis was believed to be communist-oriented. Rioting broke out on March 31, with bricks, rocks, and paint-filled eggs being thrown. The police used tear gas to break up the melee.

In rioting on April 1, nine were arrested and forty-nine hospitalized as federal troops were requested and members of Congress urged the War Department to consider taking over the plant. Representatives of both sides were invited to the President's Defense Mediation Board in Washington, DC; and, the seventy-six-day-old strike was finally settled April 6th. (Allis-Chalmers had no further strikes until after the War).

Dad probably paid little attention to what was going on in the Middle East in 1941 where the British and their Commonwealth troops attacked Iraq, overthrowing the pro-German ruler of that country in order to secure vital Persian Gulf oil supplies, including a pipeline to a refinery on the Mediterranean at Haifa in Palestine (today's Israel), where British ships refueled. The Brits also attacked Vichy French-controlled Lebanon and Syria to prevent German and Italian planes from refueling there to attack Suez and Alexandria, in Egypt—where the British had their largest Mediterranean port and airfield.

Due to the naval warfare between the Italians and British in the Mediterranean and Red Seas, the President banned US cargo ships from going there. However, in April of 1941, after the Italian Navy was defeated in East Africa, the president allowed US shipping in the Gulf of Aden and the Red Sea; but our ships were not permitted to go from the Red Sea through the Suez Canal to the Mediterranean, which was still considered too dangerous. This allowed US transports to round the Cape of Good Hope and assist the British via Red Sea harbors in Egypt (now Suez Port), where Prime Minister Churchill was sending precious tanks many thought should remain in England—still under threat of German invasion.

Although the Axis powers controlled most of Europe, the *Luftwaffe* had been unable to defeat the Royal Air Force; and the Royal Navy succeeded in imposing a shipping blockade on Nazi Germany, making Germany somewhat reliant on its new ally the Soviet Union for certain raw materials, like oil.

Since Germany and the Soviet Union had become economic and military allies in 1939, and agreed on carving up Poland, most Americans were shocked when Hitler, hoping to take advantage of a neutral USA and an isolated England, *invaded* the Soviet Union and the Soviet part of Poland in June of 1941, lured by rich Soviet raw materials, farmland,

and petroleum reserves—and based on his own personal hatred of Slavs and of what he saw as Jewish-led Bolshevism.

Surprisingly, the Soviets suddenly were *allied with Britain*.

Events in Europe were having effects in Asia, too. Japan sought to take advantage of the defeated European colonial powers, like France—occupying Vichy French Indochina (today's Vietnam, Laos, and Cambodia) to prevent English and American Lend-Lease aid from getting to China, still at war with Japan.

In July of 1941, in response to Japanese aggression, President Roosevelt issued an executive order freezing all Japanese assets in the US and banning the sale of petroleum to Japan. In London, the Dutch government-in-exile did the same for the Dutch East Indies (today's Indonesia), the largest oil producer in Asia.

With its main sources of fuel and other raw materials now cut off, Japan began making plans to secure South Asian sea routes by future attacks on Dutch, British, and American interests.

In August of 1941, the British and Soviets, now unlikely Allies, attacked Iran and overthrew its pro-Axis shah, replacing him with the young crown prince, Mohammed Reza Pahlavi, to protect Persian Gulf oil and to create the *Persian Corridor*, a system of ports, roads, rail lines, and even truck and aircraft reassembly plants, that enabled Britain, and the US, through Lend-Lease, to use Iran on the Indian Ocean, to supply the Soviets with war matériel—a much easier sea voyage than the Arctic Ocean convoys to Murmansk and Arkhangelsk. (Most Americans today are unaware 30,000 US troops served in Iran during the War. The presence of Soviet, British, and US troops in Iran was why the Big Three chose Tehran as a place to meet in November of 1943).

. . .

Tuesday, September 16, 1941—shortly after Dad started his senior year of high school—was a red-letter day for Milwaukee; and, coincidentally, as it turned out, an *unforgettable* day for Dad in West Allis.

There was no school because the 1941 National American Legion Convention was being held in Milwaukee—featuring Army Chief of Staff General George C. Marshall and entertainment by Bob Hope and

On September 16, 1941 the American Legion Convention parade on Kilbourn Avenue in downtown Milwaukee passes the venerable Republican Hotel (where, incidentally, Major League Baseball's American League had been founded 40 years before).

other Hollywood celebrities. That day (Sept. 16) had been set aside for what was the biggest parade Milwaukee had ever seen.

An estimated nine hundred thousand spectators along the three-mile parade route in downtown Milwaukee heard music from some four hundred bands, fife and drum, and bugle corps and saw baton-twirling majorettes, color guards, numerous floats, and the marching, flag-waving World War I veterans representing American Legion posts from all across the country. The parade lasted *twelve hours* and brought out a lot of patriotism in a community which, like the whole country, was torn between a desire to stay neutral and an administration's policies that seemed bent on drawing the nation closer to war.

Shortly after 9 p.m. while the big parade was ending in Milwaukee, Dad was near the West Allis A&P supermarket, located across Greenfield Avenue from the City Hall. He was in a car with his good friend John Hammer. The car was parked in front of their friend Rich Bernier's house on S. 75th Street at the entrance to an alleyway that ran along the

Photo of defendants in Stern murder trial. West Allis Star. (*Courtesy West Allis Public Library*).

side of Rich's house and behind the A&P, a small liquor store, and other Greenfield Avenue businesses.

Suddenly, the boys heard what sounded like a loud *gunshot*—apparently from Greenfield Avenue.

Seconds later, their attention was drawn to two young men in the alley, visible under the streetlight, and running in their direction from the parking lot between the A&P and the liquor store. After the suspicious teens went by, Dad and John got out of the car and rushed to Greenfield Avenue to see what had happened.

On the sidewalk, a few feet from the entrance to the liquor store, they were among the first to find its proprietor, Harry Stern, fifty, collapsed and bleeding heavily from his right chest—the obvious victim of a holdup attempt. Dad, as an employee of the A&P next door, likely knew the man, if only by sight.

With a crowd now gathering and police arriving on the scene, the unfortunate man quickly lost consciousness—and, as Dad described it to me—he literally *died in Dad's arms*.

A later published report said Mr. Stern, the married father of two daughters, was pronounced "dead on arrival" at County emergency hospital. Dad and John Hammer gave statements to police describing the assailants (who, it turned out, drove to Milwaukee afterwards to drink beer and participate in post-parade merriment).

Within a few days, an anonymous phone call led police to arrest two young Milwaukee men in their late teens who confessed to the killing and to several other robberies, mostly of drug stores and beer depots, in a crime spree that lasted months. In an attempt to avoid life imprisonment, the maximum sentence in Wisconsin, the young robbers claimed the shooting was accidental and occurred during a scuffle with Mr. Stern, who had refused to hand over any money.

In late November of 1941, when the case was tried in downtown Milwaukee, Dad was excused from school to appear as a witness—quite a memorable experience for him. I remember him telling me he was especially impressed by the prosecutor (Assistant District Attorney—later Congressman—Charles J. Kersten). Dad called him, "a real spellbinder."

The outcome was that the jury—the first *all-female* jury in a Milwaukee County criminal trial—convicted both defendants of first-degree murder. The shooting and the trial were reported in the major Milwaukee newspapers, which mentioned Dad by name, and in newspapers throughout Wisconsin—Dad's "fifteen minutes of fame."

. . .

All during the Depression, West Greenfield Avenue in West Allis, near where Dad lived, was on the route for all the parades in the suburban city, such as for holidays like Decoration Day (today's Memorial Day), the Fourth of July, Armistice Day (today's Veterans Day), and Labor Day. I remember Dad telling me, some German-Americans were proud of Hitler's accomplishments in Germany and marched in the parades in West Allis with swastika flags, or had swastika flags on their parade-cars.

At the same time, *isolationism,* the movement to keep the US out of the war, championed by the America First Committee and activists like pioneer aviator Charles Lindbergh, was very popular.

In general, most Americans had a bad taste about World War I and wanted no part of sending our soldiers to foreign lands again—instead preferring we remain neutral.

But Dad told me *all that changed with Pearl Harbor.*

3

INDUCTION

On Sunday December 7, 1941, during Dad's senior year of high school, the Japanese, in what was viewed as a sneak attack, bombed Pearl Harbor in Hawaii and American and British bases in Asia. Congress passed a Declaration of War against Japan the next day, hours after Britain did the same. The US now found itself officially in World War II.

A few days later, on December 11, 1941, honoring their *Tripartite Pact* partnership and their pre-Pearl Harbor oral promises to naval superpower Japan—Germany and Italy declared war on the United States. Within hours, the United States Congress reacted by declaring war on Germany and Italy.

With the Russian winter having halted the German juggernaut toward Moscow, and a Soviet counterattack suddenly ending Hitler's dream of conquering the Soviet capital in 1941, he welcomed the ability to wage unrestricted submarine warfare on Lend-Lease shipping and anticipated the US, now tied up in a two-front, two-ocean war, would allow him free reign to conquer the USSR in 1942—after which he would have the resources to fight the growing Anglo-American threat in the West.

Pearl Harbor had been only part of a bigger Japanese war strategy. Just before the carrier planes attacked in Hawaii, British Malaya was invaded; and *within hours* after Pearl Harbor, the strategic British colonies of Hong Kong and Singapore were bombed as well as US bases in the Philippines, Guam, and Wake Island.

After Germany and Italy declared war on the US, their Axis partners—Hungary, Romania, Bulgaria, Albania, Croatia and Slovakia—declared war on the United States. President Roosevelt tried to negotiate with the lesser Axis Partners, and the US never declared war on Albania, Croatia, or Slovakia; but, finally, Roosevelt asked for, and received, a Declaration of War from Congress on Hungary, Romania, and Bulgaria on June 4, 1942—the last time Congress declared war. (Many times since, Congress has authorized the use of military force, but has not granted the President the vast constitutional powers available under a Declaration of War).

In the weeks following Pearl Harbor, the United States went on a feverish diplomatic initiative. On January 1, 1942, the US, UK, USSR and China signed a treaty in Washington, DC, called a "Declaration by *United Nations*," in which they pledged to use their full resources against the Tripartite Pact enemies, to cooperate among themselves, and not to make any separate peace with the enemy. Twenty-two other countries signed the next day, and an additional twenty-one joined *the United Nations* by the end of the War—a total of forty-seven *countries* by 1945.

Google Maps Street View of where Dad was living during high school and afterwards. (©2017 *Google*).

FDR's secret visit to West Allis. September 19, 1942. The president, in black armband for his late mother Sara, is seated behind Wisconsin's Governor Julius Heil. (*US Navy*).[c]

When Dad, one of a class of 272, graduated from high school on June 11, 1942, at age 17—in the first graduating class after the US entered the War—he was still living with Uncle Bill and Kay and Ray in an upstairs flat at 7209 W. Madison Street in West Allis.

After graduation, if not before, Dad worked at the A&P store full-time. He may have been at work there on **September 19, 1942**, when President Roosevelt made a *visit to West Allis* as part of a secret, nation-wide, war-readiness, fact-finding tour. A switch-engine brought the president's special railroad car to a siding at the huge Allis-Chalmers factory. The president was met by Wisconsin's German-born Jewish governor, Republican Julius Heil, factory representatives, and other dignitaries and was driven around the plant. But the hush-hush visit was not reported in the wartime press until **October 1**—*twelve days later,* when the whole tour was over.

Dad had wanted to go into the service. He told me he tried to enlist in the Marines, apparently while still in high school, but he only weighed 115 lbs.; and the Marines had a 118-lb. minimum weight limit. Also—at only 17—Dad would have needed parental consent to enlist, and I assume Uncle Bill, his legal guardian, was opposed.

Under the Selective Training and Service Act of 1940, the first draft registrations were for those over the age of 21. The minimum age was then lowered to 20.

However, under the Fifth Registration for Selective Service, to be held June 30, 1942, all men *between 18 and 20* were supposed to register—that is, those born *before* June 30, 1924. To accommodate all those registering,

Milwaukee County Draft Board 28 in West Allis held the registration at the gymnasium at McKinley School, just over a block from where Dad lived, and at another school in a different part of the city. Among the 464 young men who registered for Draft Board 28 that day was Dad—*illegally*, since he was still 17 and would not turn 18 until September 11, 1942.

He never told me about this, but Dad's records show when he went to register on June 30, 1942 with friends John Hammer, Rich Bernier, and Paul Franson, only 19 days after high school graduation, he *lied about his age*—giving his year of birth as 1923, instead of 1924.

Dad was one of over 63,000 men in Wisconsin who registered that day. The 20-year olds were expected to be called up first, followed by the 19-year olds, and the 18-year olds last. Dad knew he could keep working full-time at the A&P for a while before he would actually have to appear for service. Meanwhile, Draft Board 28 moved from the police station to an office on Greenfield Avenue about a block and a half from where Dad lived.

. . .

Pearl Harbor had resulted in a great deal of patriotic fervor, and many young men—and some not so young—enlisted. Milwaukee's charismatic 34-year-old bachelor mayor, Carl F. Zeidler, who had defeated six-term Socialist mayor Dan Hoan in 1940, and who supposedly could sing in five languages, took an unpaid leave from City Hall to join the Naval Reserve as a lieutenant, junior grade. (The Common Council had refused his resignation).

In charge of Navy gunners who served on a Merchant Marine cargo ship, Mayor Zeidler ended his first sea duty in New York harbor after one of his gun crews possibly sank a surfaced German U-boat. On Sunday, August 25, 1942, colorful New York mayor Fiorello LaGuardia, in his regular Sunday morning radio "chat," revealed he had received a surprise visit from Milwaukee mayor Zeidler, just off his ship in his Navy uniform. LaGuardia said he envied his Milwaukee counterpart, who now no longer had all the troubles of a mayor—though facing the danger of enemy submarines.

Zeidler put in for a seventy-two-hour furlough. After midnight on August 26, he landed at the Chicago Municipal Airport (not yet called

Mayor Carl F. Zeidler.[c]

Midway) where his parents and a city driver picked him up and drove him to Milwaukee for a whirlwind visit. After a few hours rest, he was welcomed at his old office at City Hall by Common Council President/Acting-Mayor John L. Bohn and was overwhelmed by city employees and other well-wishers.

Local newspapers announced the mayor was unexpectedly back in town—if only for a very brief visit.

The next day, August 27, the traditional "Milwaukee Day" at the State Fair in West Allis was suddenly renamed "Zeidler Day" as the six-foot-two, blond, blue-eyed naval officer/mayor accompanied Acting-Mayor Bohn for a tour of the Fair. Zeidler was swamped by his many fans who followed him around as he did such things as meet with Sea Scouts and Boy Scouts, sample famous State Fair cream puffs, and sign as many autographs as time allowed. At the grandstand, he addressed a crowd of about fifteen thousand, saying "Let's put an ax to the Axis!" After leaving West Allis, he made more public appearances in Milwaukee.

The following day, he left Milwaukee for Chicago and the return flight to New York—taking leave of his parents, sister, and two brothers. Back in New York, at a Navy-organized press conference, he praised the civilian sailors on cargo ships and said members of the Merchant Marine should have uniforms and be allowed to visit USO and other service clubs, like members of the military. He wrote a letter to his parents in Milwaukee, dated September 9, 1942.

It was the last they ever heard from him.

. . .

On December 15, a postal telegram from the Navy was delivered to Mr. and Mrs. Zeidler regretting to inform them their son, Lieutenant [j.g.] Carl F. Zeidler, USNR, was officially *missing in action*. His ship, it turned out, had disappeared at sea.

Milwaukeeans were in shock. Their *golden boy*, the youngest mayor of any major American city, was expected someday to become governor, or a US Senator. The Navy, after two years, officially declared Mayor Zeidler dead; but his parents, family, and friends still held out hope he was somehow alive. Despite their numerous inquiries and requests by prominent Wisconsin politicians, the Zeidler family did not learn what actually happened until decades later—with the help of German records.

At a time of heavy U-boat activity in the Atlantic, Mayor Zeidler had been in charge of about twenty Armed Guard Navy personnel aboard an old, armed—but unescorted—merchant steamship that left New York September 10, 1942, proceeded to Guantanamo, Cuba, went through the Panama Canal, and supposedly was last seen on September 26, 1942, at Balboa in the Canal Zone, near Panama City, on the Pacific side of the canal. The S.S. La Salle, under radio silence and following a diversionary route, was carrying Lend-Lease trucks, aircraft parts, and ammunition destined for Karachi—then in British India—a port in the China-Burma-India Theater of the War.

After steaming down the West Coast of South America and Cape Horn, the vessel continued across the South Atlantic toward the Indian Ocean. After dark on November 7, 1942, the La Salle was torpedoed by a surfaced German U-boat over three hundred miles southeast of Africa's Cape of Good Hope. The merchant vessel's crew and Navy complement had no chance as the ship's ammunition cargo violently exploded in a tremendous fireball—falling debris even injuring several German sailors on the conning tower of the surfaced U-boat.

Sadly, all sixty aboard the La Salle, including Milwaukee Mayor Zeidler, were killed.

. . .

While awaiting his draft notice, Dad, I assume, did his best to enjoy himself during the summer and fall of 1942. He had graduated from high school and had money in his pocket from his job at A&P. He and his friends, all awaiting an uncertain future, undoubtedly hung out together—while they could. I am sure Dad had an active social life. He started a photo scrapbook while in high school and another while in service. He never spoke of his premarital social life, but the scrapbooks contained photos of a Phillis, Lil, Corine, Marion, Ada, Irene, and a Laura.

On December 18, 1942, a notice was mailed to Dad to appear at the Selective Service board, and he presented himself on December 21. His occupation was "grocery clerk at A&P" and his duties involved handling cash, working on stock, and building displays. He was making $22 per week and had worked there over a year. The following day, he was transported to the Induction Station in Milwaukee and given a pre-induction physical examination.

On the radio that holiday season was a popular song written by Irving Berlin, recorded earlier in the year by Bing Crosby for a movie called *Holiday Inn*. The song was called "White Christmas." It has been played on the radio every holiday season since, and became the best-selling song ever recorded.

On December 28, 1942, right after Christmas, Dad was classified 1-A, the top classification for being drafted.

On January 18, 1943, as ordered, Dad appeared at the draft board and was taken by bus to the Milwaukee Induction Station, which I understand was on Edison Street in the warehouse area near the Milwaukee River, for a further physical and for transportation. After he passed the physical, he met with a special board regarding which branch of service he preferred; and, as part of a group, he was sworn in. Technically, his status was "inactive" until he was actually accepted by one of

Dad's 1942 graduation picture.[c]

the branches of the armed forces, in his case the Army. Still in civilian clothes, he was taken by bus or train to the nearest U.S. Army Reception Center, which was at Fort Sheridan, Illinois, on Lake Michigan, north of Chicago.

In his ten days at Fort Sheridan, he was fingerprinted, fitted for fatigues, boots, shoes, and a dress uniform, attended classes, was given vaccinations, got a haircut, endured close-order drilling, learned basic military courtesy and discipline—and, like most GIs, spent lots of time standing in line. He had another physical and was found to be 5'8" tall, weighing 123 lb., with brown hair and brown eyes, and blood type B+. A chest x-ray was negative. Dad had to undergo a number of written intelligence and aptitude tests and had a one-on-one session regarding what he would be best suited to do in the military—information put on his WD AGO Form 20 (not one of the records that survived). His civilian occupation was as a grocery clerk and window trimmer making $22 per week. His job had been to deal with customers and set floor and window displays. He gave his mother as next of kin; and he designated Uncle Bill his beneficiary and Kay's husband, Ray Blake, as an alternate beneficiary. His age was given as 19 years, 4 months (whereas really, of course, he was 18 years, 4 months).

Because he had lied about his age when he registered for Selective Service, Dad's date of birth, throughout his military records, was always given as September 11, *1923*—instead of 1924. As far as the Army was concerned, Dad was always *one year older than he really was.*

Like most recruits, Dad probably made the common mistake at Fort Sheridan of addressing his drill sergeant as, "Sir," only to be told, "'Sir' is for officers—*I work for a living!*"

While in the service, Dad, who had spent his whole life around Milwaukee, was often asked where he was from; and he would say, "West Allis." He told me this often prompted a drawling response from a fellow soldier like, "Oh, I'm from *Texas,* too!" And Dad would then have to explain he was not from "West Dallas" but from a city in Wisconsin.

Three days later, Dad, who had never been farther from home than Chicago, was sent for basic training on the *longest train ride of his life* to Camp Davis in southeast **North Carolina**, to a unit called Battery D, 391st Coast Artillery—and he arrived January 30, 1943.

4

CAMP DAVIS

Virtually nothing remains of Camp Davis, North Carolina today, but when Dad arrived there as a buck private for basic training at the end of January in 1943 it housed about 20,000 Army personnel, mostly in two-story wooden barracks.

Since the War of 1812 and the defense of Fort McHenry at Baltimore harbor, which gave us our national anthem, the United States appreciated the importance of using long-range coastal guns to protect America's ports from enemy ships. Unfinished and undermanned Fort Sumter in Charleston harbor, South Carolina, was shelled from the mainland by the Confederates in 1861, triggering the Civil War.

In 1907, Congress made the Coast Artillery a separate branch of the Army from the Field Artillery. However, even before Pearl Harbor, it was clear the biggest threat to the homeland would come from the sky rather than the sea, and Camp Davis was envisioned as a major anti-aircraft artillery training center.

So, in 1941, Camp Davis, comprised of some 3,000 buildings on

Camp Davis, North Carolina.

about 50,000 leased acres of land, about 30 miles northeast of Wilmington, North Carolina, was built *in less than six months*.

Shortly after Dad arrived for basic training, his battalion lost its "Coast Artillery" designation and was re-named the *391st Anti-Aircraft Artillery Automatic Weapons Battalion*.

Dad's rise among the ranks, whether by luck, talent, or what some would call *chutzpah*—even for wartime—was amazing.

This is the story as I recall Dad telling it. Shortly after he arrived at Camp Davis, Dad's basic training commanding officer, probably a captain, spoke to his troops and said if any of them had a complaint, to come and see him. I'll be honest. Dad was always a very fussy eater, a meat-and-potatoes guy, not given to variety, and probably not too partial to some of the Southern food served in the Army. Anyway, he went to the commanding officer and said he had a complaint. *What was it?* he was asked. "The food tastes like *shit*, sir," Dad replied. As luck would have it, the mess sergeant—in those days of food-rationing—had just confessed to selling government-issue food on the black market and had been busted. Dad's commanding officer asked Dad, since he showed such interest, if he had any experience with food. Dad, turning on his street charm, said he had worked in a grocery store and *knew all about food*. The commanding officer told Dad he now had an opening for a mess sergeant; and there were certain basic-training classes he could not get him out of; but, otherwise, Dad was to run the mess hall—and if all went well, Dad would make rank all the way up to staff sergeant, the rank for a mess sergeant, in the absolute minimum amount of time.

And that is what happened. According to the individual pay record he kept, Dad made *Private First Class* on March 16, 1943, *Corporal* on March 27, *Sergeant* on April 15, and *Staff Sergeant* (the equivalent of today's E-6) on June 27—only five months after going on active duty, and still only age 18. His base pay was increased to $96 per month—more than he had made at A&P as a civilian.

He remained a staff sergeant the rest of his time in service. Dad, who had *worked* during school, and was not in scouting, team sports, or extracurricular activities, was—it turned out—a natural leader.

As a mess sergeant, Dad was responsible for feeding all the officers and enlisted men in the battery. A battery in the artillery is equivalent to a company in the other Army branches, usually about a hundred men. Dad was responsible to a lieutenant in the battery designated—among his many duties—as the *mess officer*. Otherwise, Dad ran the mess hall. He had to be familiar with nutrition, menu-making, sanitation, rations and mess accounts and books. Cooks worked under him, as well as all the enlisted men peeling potatoes on KP (kitchen police) duty, serving food in the chow line, and scouring dirty pots and pans. He decided, with the mess officer, what the men would eat, ordered the food from the quartermaster corps, and was responsible for all the food storage and preparation and the cleanup afterwards. He had responsibility for

Dad as an eighteen-year old corporal in North Carolina.[c]

Staff Sergeant Finney with Private John Hammer (standing) and Corporal Jack Jung, both of whom later would also become sergeants. (West Allis. July 1943).

all the equipment used in the mess hall and assisted in keeping the books and records necessary to account for ordering and budgeting. Dad had to get up early at Camp Davis. Reveille was at 5:45 a.m., by which time Dad and his crew were already in the process of preparing for breakfast at 6:30 a.m. In the field, Dad had to set up a field kitchen. If he could not do that, he had to distribute C-or K-rations or otherwise see to it everyone in the battery got fed. It was a big responsibility for an eighteen-year old former grocery clerk, but Dad did a good job and got promotion after promotion.

In late July of 1943, Dad, always a bit cocky and quite proud of his rapid advancement, came home on ten days leave, his second furlough, and got together with his buddies John Hammer and Jack Jung, who were also in the service—and whom he now outranked.

At 7209 W. Madison Street, Dad's "home" address, Kay and Ray had a daughter, Jerilynn, called "Jill"; and Dad got to see his little second-cousin during his leave in July of 1943.

After basic training, Dad did not move on with his unit. At the beginning of July, before he went on leave, Dad was selected for what was called a "cadre pool," meaning he would be reassigned to a new training unit that needed a mess sergeant, but would stay at Camp Davis. He was assigned on July 8, 1943, to Battery D of the 557th Anti-Aircraft Artillery Battalion.

Even before the U.S. entered World War II, the War Department appreciated that our Air Force, then part of the Army, was inferior to the German *Luftwaffe* in pilots, tactics, and in the number and quality of warplanes. The speed and accuracy of *Stuka* dive bombers in advance of Panzer tanks had created a new kind of mobile warfare called *Blitzkrieg*, or "lightning war," first seen in Poland in 1939 and to devastating effect against the French and English in 1940. In the Pacific, we had no warplane to match the Japanese *Zero*. It was clear an Allied victory would require a massive buildup of aircraft and the training of pilots, flight crews, and ground personnel. Until that could be accomplished, and air superiority could be achieved, the US would need to rely on anti-aircraft fire to protect not only harbors on our soil, but also our ships, ports, airfields and troops overseas. That is why anti-aircraft artillery became a priority, leading to the establishment, almost overnight, of some seven anti-aircraft artillery training centers, like Camp Davis, where Dad was stationed.

Dad, on leave, with Kay and Ray's daughter Jill. (Note "AA" patch for *Anti-Aircraft*).c

Ray, Kay, and Jill Blake.[c]

Dad facing the heat. Probably at Fort Fisher.

While technically established as part of the *Coast* Artillery, the sprawling Camp Davis was actually a few miles *inland* from the coast, near what had been a railroad refueling stop called Holly Ridge, NC which had a population of under fifty people in 1940. The camp, which had its own airfield, also operated five coastal firing ranges, the biggest of which was at the site of an old Confederate Civil War fort called Fort Fisher, about twenty miles south of the military town of Wilmington, NC, and fifty miles from Camp Davis. Like Camp Davis, Fort Fisher was hot, humid, and loaded with mosquitoes. But it was also on the Atlantic Ocean with a sandy beach—and water a lot warmer than at the Lake Michigan beaches in Milwaukee Dad was familiar with.

During the Civil War, Fort Fisher, "the Gibraltar of the South," had protected blockade runners going in and out of Wilmington from Bermuda and the Caribbean. Wilmington, on the Cape Fear River, had been the last Confederate port on the Atlantic to fall. (The capture of Fort Fisher and Wilmington in early 1865 by Union forces deprived Confederate General Robert E. Lee's army of a major source of supply and contributed to Lee's surrender a few months later).

On September 2, 2020, President Donald Trump named Wilmington—which Dad likely visited often—America's first **"World War II Heritage City"** for its contribution to the war effort.

Although some gun crews from Camp Davis practiced live-fire with 90mm guns intended for *high-flying* enemy bombers, (many of which guns had Allis-Chalmers gun carriages), Dad's units, known as *automatic*

The Bofors L60 40mm anti-aircraft gun.[c]

Typical mess at Camp Davis in 1942 (before Dad arrived). (*New Hanover County Library*).

weapons battalions, trained on what was known as the 40mm Swedish-designed L60 *Bofors* anti-aircraft gun, which could fire up to 120 rounds per minute, was effective against *low-flying* strafers and dive bombers, and came in a variety of configurations—towed, mounted on a truck, air-cooled and water-cooled.

Virtually every Navy ship was armed with Bofors guns, usually in twin or quad configurations. Some sixty thousand of them were manufactured by Chrysler Corporation. Army Bofors guns were the primary armament of *automatic weapons* battalions, like Dad's, which also were equipped with .50-caliber heavy machine guns, often quad-mounted.

During the First World War, for the primitive, static-filled radios of the time, the British developed a phonetic alphabet that used *words* as a substitution for letters. For the letter "A" they used the word "ack." Hence "AA," or anti-aircraft fire, was called "ack-ack." The term stuck and was adopted by the Americans and used in World War II, as well. Soldiers in anti-aircraft units were sometimes called "ack-ackers."

To free up male pilots, a group of women fliers, called WASPs—Women's Army Service Pilots—took on the dangerous role of towing targets from the Camp Davis airfield to the coastal firing ranges, like Fort Fisher, so the soldiers could train with live ammunition. Besides anti-aircraft artillery and heavy machine guns, training for some Army units at Camp Davis was on radar and searchlights.

As part of the Camp Davis cadre pool, Dad, a mess sergeant, now training others, kept getting reassigned every month, or so, to a different

A Bofors during training at the beach firing range at Fort Fisher, North Carolina.[c]

training unit. From September 25 to October 24, he was with Battery D of the 445th Anti-Aircraft Artillery Battalion.

From October 25 to November 19, Dad was with Battery D of the 446th Anti-Aircraft Artillery Battalion, his last assignment at Camp Davis, after which he had a ten-day leave and presumably went home.

On December 1, after his Thanksgiving leave, he reported to Battery B of the 589th Anti-Aircraft Artillery Automatic Weapons Battalion, at Camp (now Fort) Stewart, Georgia, an hour southwest of Savannah. Dad only spent a few weeks there. He got a Christmas leave starting December 23, 1943, his last leave before going overseas.

I suspect coming home on leave was quite an adjustment for Dad, even changing how he talked. I am sure he learned *profanity* long before going into the Army; and, like most GIs, he accepted swearing as a part of Army life. He often used the expression "swore like a trooper" to describe someone who used profanity.

At home on leave, of course, he had to clean up his act a bit. Movies from the 1940s never had any profanity in them because of the Motion Picture Production Code. Dad told me no movies or even books of the time really captured how soldiers actually spoke. It was not until I was in the Army myself, that I appreciated how much profanity was a part of everyday speech in the military.

The subject of profanity reminds me of a famous bar in Milwaukee Dad told me about visiting, probably after the War, known as "Dirty Helen's." I discovered the actual name of the bar was the "Sunflower Inn" on W. St. Paul Avenue run by a former prostitute and madam named Helen Cromwell, aka *Dirty Helen*. It had been a speakeasy before Prohibition ended, so Dad probably heard about the place from Uncle Bill. Helen's bar sold only scotch and bourbon, and the only mix was water. Order anything else, and Helen, usually tending bar, was going to call you every four-letter word in the book, much to the amusement of the other customers. It was said Dirty Helen could out-swear any man. She became a Milwaukee institution.

While on leave, Dad was aware of the sacrifices being made on the home front. With factories devoted to wartime production, no cars were manufactured. Speed limits were lowered to 35 mph, and gasoline was rationed, as were tires due to a rubber shortage. To save copper, pennies were made of steel. Recycling of cans and scrap metal was encouraged. Homeowners were urged to grow their own vegetables in "victory gardens." It seemed like every movie star was selling war bonds. The movie

An Allis-Chalmers shop making propeller shafts for Navy warships during World War II. (*West Allis Historical Society*).

theaters all sold war bonds and stamps. Women were not only doing traditional clerical jobs at factories like Allis-Chalmers—now with twenty thousand West Allis employees—they were also operating drill presses and punch presses, doing assembly on production lines, and soldering and welding components. Allis-Chalmers even built a separate plant in next-door West Milwaukee to manufacture General Electric-designed turbo superchargers that enabled the piston engines of our high-altitude bombers to operate in the thin air of the upper atmosphere. Four out of five employees at the supercharger plant were women.

Due to its wartime production, Allis-Chalmers was given the Army-Navy "E" Award by the War Department. The "E" was for "excellence" in the quantity and quality of its production. The factory was given a special pennant to fly on its flagpole, and each worker was given a pin commemorating the event. Of the over eighty-five thousand factories involved in wartime production, only about five percent earned the Army-Navy "E" Award.

Groceries during the War were rationed by the Office of Price Administration. Every American family was registered, and ration stamps were sent through the mail on a point system with each person in the family allowed so many points per week. Blue Stamp rationing covered processed foods like soups, condiments, juices, and canned fruits and vegetables. Red Stamp rationing covered meats, dairy and vegetable oils, among other things. Rationing also involved items like clothing, shoes, coffee, gasoline, and fuel oil. The shortages of automobiles and gasoline meant most people used mass transit. The Milwaukee area's orange-and-cream buses and streetcars were never so crowded as during the War.

I know Dad appreciated how the American people were willing to sacrifice so soldiers like him and the men in his mess could be well-fed and well-supplied. However, even soldiers had to be mindful of how food was managed. Camp Davis was particularly concerned with limiting waste from the mess halls. *Take all you want, but eat all you take* was the rule.

. . .

Bing Crosby had another hit song for Christmas of 1943. It was called "I'll Be Home for Christmas (If Only in My Dreams)" and was especially popular with the many GIs now serving in the military. With so many men at forts, camps, and bases all over the country and in many foreign lands, Dad was lucky to be home for another Christmas; but he no doubt wondered where he would spend the *next* Christmas.

When Dad left West Allis this time, he was unaware he would not be back home for nearly *two years*.

Thanks to the Army, when it was time for Dad to leave, the trains were running. Three major unions had threatened a rail strike starting December 30, 1943, so President Roosevelt on December 27 invoked his authority as commander-in-chief to seize the nation's railroads—as President Wilson had done during World War I—and the Army Transportation Corps took over. (By January 18, 1944, the unions settled, and the railroads were returned to civilian control).

A few months after Pearl Harbor, Army Chief of Staff General George C. Marshall totally reorganized the United States Army, which also included the Air Force. He created three separate "forces." The **Air Forces**, successor to the Army Air Corps, were for just about everything having to do with aircraft. **Ground Forces** were the *combat arms* like infantry, armor, combat engineering, and artillery (including anti-aircraft artillery—like what Dad was in). **Service Forces** included just about everything else, from personnel and finance to ordnance, quartermaster, medical, etc.

On January 5, 1944, Dad, being in the Ground Forces, reported, as ordered, to the *Ground Forces Replacement Depot #1* at gigantic Fort Meade, Maryland which was in charge of individual replacement manpower for Ground Forces deployed in Europe and North Africa. Dad was not to be sent overseas as part of a unit, but as an *individual* replacement.

With Fort Meade being just outside of Washington, DC, Dad told me he was given a tour of the nation's capital by a female relative or acquaintance of Uncle Bill's who worked in Washington during the War and had access to a car.

I should say here that if Dad had a camera, I suspect he would have taken pictures. In fact, we do not have too many pictures of Dad's military experience, though I have included most of them in this work. I do not know if he owned a camera while in the service, or if all the pictures he brought back were given to him. I have only one roll of negatives. I suspect he did not have a camera but may have borrowed one on occasion.

From Fort Meade, Dad was sent on January 14, 1944, almost certainly by train, to the Army's Camp Patrick Henry, part of the sprawling *Hampton Roads Port of Embarkation* facility at Newport News, Virginia, near the mouth of the James River, where it emptied into Chesapeake Bay, to await transport. He was on "casual detail" there until he boarded ship on February 2, 1944. He likely was not told where he was headed until he was out at sea, if then.

He told me he crossed the Atlantic on a *Liberty ship*—the specific one I do not know. (In 1951, the Army destroyed all World War II troop ship passenger lists and manifests).

When President Roosevelt saw the blueprints for the Maritime Commission's first Liberty ships, he remarked they would do the job but were not much to look at. "A real ugly duckling," he supposedly said. From then on, the Liberty ships, like the one Dad was on, were sometimes called "ugly ducklings."

I believe Dad's vessel was part of an eighty-nine-ship, "slow-moving" convoy, organized by the War Shipping Administration, called **UGS-32**, that left Chesapeake Bay on February 3, 1944, bound for the Mediterranean Sea and various ports in North Africa.

Under the convoy system, slow cargo and troop vessels, unable to outrun German U-boats, were protected by an escort of a dozen or more Navy warships. Some ninety-five such slow convoys—averaging over seventy ships each—sailed from Chesapeake Bay to the Mediterranean during the War. Faster cargo vessels, such as Victory ships, sailed in fast convoys. The largest North American convoys sailed the Atlantic not only from Chesapeake Bay, but also from New York harbor and Halifax, Nova Scotia.

Erroneously, the *Arnold Hague Convoy Data-base* lists UGS-32, the slow convoy with Dad's Liberty ship, as "unescorted." However, a declassified log from the escort destroyer USS J.R.Y. Blakely (DE-140) proves Dad's convoy actually *was* escorted through the middle Atlantic by Navy warships from Escort Divisions 9 and 46 of Task Force 65 as far as Gibraltar. On February 19, 1944, Task Force 65 was relieved of its escort duty by a British escort from Gibraltar for the Mediterranean part of the journey.

Protection for Dad's convoy came not only from surface escorts but also from special sub-hunting aircraft. In the fall of 1943, Winston

Churchill was able to invoke a mutual-defense treaty with neutral Portugal—*from the year 1373, no less*—that allowed Britain to build airfields on the critical Portuguese Azores Island chain in the Eastern Atlantic. The British airfield in the Azores, and a subsequent larger one built by the Americans, together with an American airfield built on Bermuda in the Western Atlantic, allowed the Allies by 1944 to provide land-based air cover against U-boats for convoys, like Dad's, sailing from the southern US toward Gibraltar and the Mediterranean.

The Allies had used technology to win the so-called *Battle of the Atlantic* and minimize the U-boat threat. Long-range, land-based bombers, like B-24 Liberators, were fitted with microwave radar, improved altimeters and navigation aids, sonar buoys, airborne depth charges, and magnetic anomaly detectors that enabled the planes to find submarines underwater, and also at night when the U-boats surfaced to recharge their batteries and travel faster under diesel rather than battery power. The breaking of German naval codes and the decryption of the Enigma code machine brought intelligence into the picture. Particularly successful was a campaign to destroy special refueling U-boats, called "*Milch* Cows." The air cover got so good the very presence of aircraft in the sky caused many U-boats to forego attacking.

In addition to land-based aircraft, convoys were often protected by *escort carriers*, sometimes called "baby flat-tops," about half the length of a regular fleet carrier, and slower—but ideal for Navy scout planes, mine layers, and sub-killers. Escort carriers were also used to support land troops in amphibious assaults, especially in the Pacific. (In October of 1944, crewman Vernon Ponto, father of my brother-in-law Steven Ponto, was at the Battle of Leyte Gulf, aboard the escort carrier USS *St. Lo* (CVE-63) when it became the first major warship sunk by a *kamikaze* attack).

As nearly as I can tell, although U-boats posed a threat until the end of the War, when Dad crossed the Atlantic in February of 1944, the few remaining U-boats were mostly interested in attacking ships destined for Britain during the Normandy invasion buildup—and Dad's more southern crossing was uneventful.

5

NORTH AFRICA

I am sure Dad had never before been on an ocean-going vessel. The Liberty ships had a top speed of only 11 knots, but usually cruised at 10 or less; and the convoy could not go faster than the slowest ship. Dad told me one of the things the GIs did to pass the time was to have a seasick competition. The participants would pay an entry fee into a pot, lie down on the rocking deck and watch a tall mast sway with the waves. The rule was they had to keep their eyes open and on the swaying mast. One after the other, the guys would get nauseous and have to go to the rail. After a set amount of time, the ones who did not get seasick split the pot of money. I don't recall if Dad was one of the winners, but I doubt it.

Dad told me another anecdote. At some point several GIs went swimming off a raft behind the Liberty ship. Later, they reported to sick bay with bad sunburn and were *busted*. What for? Dad told me a soldier was considered government property, and getting sunburned in the Army was considered *damaging government property*.

About 2,700 Liberty ships were manufactured, many built during the War in less than a month. They were designed by the Maritime

An "Ugly Duckling" World War II Liberty ship. (*U.S. Navy*).[c]

Commission in the mid 1930s to carry cargo. However, starting in September 1943, on an emergency basis, when ocean liner transport became inadequate, about 225 Liberty ships, like Dad's, were fitted as *troop transports* and could carry 350-550 men. One of 18 different manufacturers of the steam engines used for Liberty ship propulsion was Allis-Chalmers' Canadian subsidiary.

Allis-Chalmers was also a major supplier of the steam turbine propulsion engines used by the larger and faster merchant marine *Victory* ships, of which over 500 were manufactured during the war, and of the turbines used to generate electricity aboard the vessels.

Armed Liberty ships, like the one Dad went overseas on, were crewed by merchant marine sailors, but had twenty to forty US Navy Armed Guard personnel on board for anti-aircraft and anti-submarine defense of the vessel.

Dad actually received a *commendation* from Navy Lieutenant R. B. Reid, USNR, for his service as a volunteer gunner and lookout on his Liberty ship during the crossing.

On February 20, 1944, *after eighteen days at sea*, Dad's Liberty ship troop transport, along with about six others and several cargo ships, docked at Oran, Algeria, at the foot of the Atlas Mountains in North Africa. Oran provided logistical support for Allied operations in the

Mediterranean, including the invasions of Sicily and Italy and the planned invasion of Southern France.

On July 3, 1940, two weeks after France had surrendered to Germany, Oran's nearby port at Mers-el-Kébir, had been the scene of an attack on French warships *by the British Navy,* fearful of the threat posed if the Vichy French fleet—fourth largest in the world and dominant naval power in the Mediterranean—were combined with the German and Italian. Although, under terms of the French-German armistice, the French Navy was to remain neutral—and Vichy French Admiral François Darlan had promised Winston Churchill no French warships would fall into German hands—Churchill, nevertheless, made the decision to attack the part of the French fleet then in port at Mers-el-Kébir. Nearly 1,300 French sailors were killed in the controversial attack by their former ally, which poisoned British-French relations. Most of the French Navy's other capital ships remained in the Mediterranean port of Toulon in France. President Franklin Roosevelt applauded the attack, both because it somewhat removed a threat to US interests and because it showed Britain's resolve under its new prime minister, in office only two months, to fight—whatever the consequences.

Oran fell into Allied hands in November of 1942, when the then Vichy-controlled French offered only a few days of resistance to a combined British-American invasion called *Operation Torch.* Six months later, combined British, American, and Free French forces totally defeated the Germans and Italians in the Battle of Tunisia, bringing all of North Africa under Allied control.

When Dad arrived in Oran, the fighting in North Africa had been over for nine months, and the Allies had taken Sicily and had begun the invasion up the boot of Italy, securing the critical deep-water port of Naples, which took several months to clear of scuttled ships, mines, debris, and other hazards left by the Germans who were establishing defensive lines between Naples and Rome.

Like his father before him when seeing Santiago de Cuba, Dad must have been shocked to see wartime Oran as he was trying to get his "land legs," coming down the gangway under his heavy duffel bag.

In his memoir *Granddaddy, Tell Us About the War,* author Kenneth K. Gowen, who arrived in Oran about five months before Dad, described what he saw on a truck ride from the dock as a "filthy, dirty, stinking city." Especially disturbing were the barefoot children in ragged clothes, urinating publicly in the alleyways.

Dad was assigned to the 390th Replacement Company in Oran. Another entry put him in the headquarters company of the 8th Replacement Depot and in the "Lion Mountain Area." Lion Mountain (Djebel Khar) is eight miles northeast of Oran and at about 2,100 ft. in elevation. I have found no evidence of any other assignment before Dad left North Africa three months later.

One *plus* for Dad in North Africa was that he would now receive foreign duty pay—a raise of 20 percent. (Congress did not authorize "combat" or "hazardous duty" pay until the Korean War).

Dad mentioned to me he spent time in the Atlas Mountains, which would make sense if he were temporarily assigned as a mess sergeant for an anti-aircraft artillery battery, since such batteries were in the mountains around Oran to protect against land-based enemy aircraft still operating in the Mediterranean from France and Italy.

It was in the Atlas Mountains that Dad experienced, for the first time, one of the most painful afflictions known to man, or woman. He woke up in the middle of the night with a killer ache in his lower back—the worst pain he had ever suffered in his life. He was evacuated to an aid station below; but by the time he got there, the pain was gone, and he was fine. A few years later in Milwaukee he was moving a mattress and had a similar attack—killer pain in the lower back. X-rays and a urine test showed the culprit was a *kidney stone*. While in North Africa, he had passed the stone without realizing it. During his lifetime, Dad was to have many more kidney stone attacks and numerous urinary tract infections. Several times, I had to drive him to the emergency room for morphine. He eventually lost a kidney. The surgeon at the VA hospital who removed it told us the kidney was "like a bean bag."

Dad also told me of something that happened in North Africa that I think of as the *Ugly American Incident*. Dad was riding up-front out in the country in some kind of vehicle, probably the typical two-and-a-half-ton truck. An Arab was in the road for some reason, and the driver, instead of stopping or trying to go around the Arab, simply ran right over him, grumbling, "Goddam Arabs!"—and just kept going. I think

The knife Dad brought home from North Africa.

Dad told me this story to show how some individuals used the War as an excuse to indulge in their baser instincts.

Although the ground fighting in North Africa was over by the time Dad got there, Oran was one of several major ports participating in the buildup of men and matériel for a controversial invasion of Southern France that was originally supposed to take place in June, 1944, and complement the Normandy Invasion in Northern France—hopefully to keep the Germans from reinforcing their troops near Normandy—but it got postponed to August. A Free French and US operation, opposed by Churchill, the invasion of Southern France was to come not only from Oran but also Corsica—Dad's next duty station—as well as Naples and Taranto in Italy; Palermo, Sicily; and from the island of Malta, just south of Sicily. As we shall see, the planning and buildup for that invasion, including pre-invasion bombing campaigns in Southern France, were to have an impact on Dad's assignments in Corsica and Italy.

Dad managed to acquire a souvenir in North Africa. He bought from, or traded with, a well-to-do Arab for a knife with a decorative blade and a camel-bone handle. I have the knife. On the handle was engraved "Oran Africa 44."

On May 25, 1944, Dad left North Africa, probably from Oran. From what I remember him telling me, he went on a Free French ship. After Germany defeated France in the summer of 1940, the Germans, wanting to save their manpower for the conquest of the Soviet Union, only occupied the northern and western parts of France. When the English attacked the French fleet in Oran, about two hundred French ships were in English ports. The supposedly neutral Vichy French government, which

I think this is the Arab Dad got the knife from, April 19, 1944. (Dad's comment on the back of the photo: "This guy has more wives than I have fingers or toes!").[c]

answered to the Germans, controlled southern France and Corsica, as well as the world-wide French colonies—including Morocco, Algeria, Tunisia, and Senegal in Africa; and it kept the remnants of the Mediterranean French navy in Toulon.

When the Americans and British landed in western North Africa (Morocco and Algeria) in November of 1942, and the Vichy military appeared to collaborate with the Allies, the Germans and Italians took over occupation of the rest of France and Corsica and tried to take over the Vichy navy in the Mediterranean port of Toulon, France. Two years before, Vichy French Admiral Darlan had promised British Prime-Minister Winston Churchill (and the Americans) no French warships would fall into German hands. Following Darlan's orders, the Vichy French fleet in Toulon was scuttled before the Germans arrived.

Like elements of the Vichy army in North Africa, the remaining ships of the Vichy navy had rallied around a French general named Charles de Gaulle, who had fled France for Britain, planned to free France with troops from French colonies, along with British and American aid, and who called his government-in-exile *Free France*.

6

CORSICA

Dad did not have much good to say about the Free French ship he boarded in North Africa, calling it "filthy." On May 27, 1944, two days after leaving the "Free French" colony of Algeria, he arrived at the port of Ajaccio on the West Coast of the "Free French" island of Corsica, west of Italy and southeast of France. Corsica was where Napoleon Bonaparte had been born. Though officially French, the people there spoke a language closer to Italian. Several months after Italy switched sides in September of 1943, Corsica was freed from the occupying Axis troops by Allied-armed Free French forces (mostly Moroccans) and a well-organized resistance movement. It was the first French *Département* to be liberated.

But the island was still within range of the *Luftwaffe*, the Nazi air force.

In fact, in the evening hours of May 12 and early morning hours of May 13, 1944—less than two weeks before Dad left Algeria—the *Luftwaffe*, flying literally *below radar* from bases in Northern Italy, had carried out devastating surprise raids in the dark on two Allied airfields on Corsica. One was home to British Spitfires and the other, *Alesani*, was where Dad was destined to be stationed. The raid at Alesani destroyed or damaged sixty-five Allied aircraft, mostly medium (i.e. two-engine) bombers, killed twenty-one, and wounded seventy. The attackers did extensive strafing on anti-aircraft batteries around the airfield. I suspect the damage done by the surprise raids, injury to personnel, and the need to beef up defenses was why Dad, a replacement mess sergeant in anti-aircraft artillery, got orders for Corsica.

After his arrival at the port of Ajaccio in western Corsica, Dad was assigned briefly to headquarters battery of the 213th Anti-Aircraft Artillery Regiment, and then to Battery D of the 454th Anti-Aircraft Artillery Automatic Weapons Battalion, stationed in the eastern part of the island near a place called Alistro, on a ridge overlooking the recently-raided *Alesani* (Corsican) or *Alesan* (French) Airfield. Though mostly mountainous, Corsica had a flat lowland along its Eastern Coast, which made it ideal for Allied air operations against Italy and Southern France, and the Allies had constructed a number of temporary airfields there.

The mission of Dad's unit was to protect the Alesani airfield. I remember Dad telling me, that from Corsica, with field glasses, he could see German aircraft taking off from the Italian island of Elba to the east of Corsica. Elba, famously, was where Napoleon was exiled, the first time.

Within three weeks of Dad's arrival on Corsica, a 270-ship US and British armada, supported by British bombers, landed a Free French infantry division and commando battalion, as well as Royal Navy commandos, and captured Elba and its German airfield in three days.

The 340th Bombardment Group had only moved to Alesani Airfield in April 1944 (a month before Dad arrived on Corsica). A month earlier, it had been stationed at Pompeii, near Naples, on the Italian mainland and was devastated by an eruption of Mt. Vesuvius, losing more of its B-25 Mitchell bombers to the volcano than to the devastating German air raid at Alesani two months later.

Only a few weeks before Dad arrived at Alistro, a young lieutenant bombardier started flying on 340th Bombardment Group missions out of Alesani. His name was Joseph Heller and he went on to write about his wartime experiences in the classic 1961 novel *Catch-22*.

Dad's anti-aircraft artillery unit was defending Alesani for Twelfth Air Force that launched planes to Italy in support of what became known

Damage done to Alesani—also called Alesan—Airfield, Corsica by *Luftwaffe* raid in early morning hours of May 13, 1944, less than two weeks before Dad arrived. B-25s of the 489th Bombardment Squadron, 340th Bombardment Group. (*John Sterrett*).[9]

as the *Rome-Arno Campaign*, which went from the German *Gustav Line*, south of Rome, in January, 1944—before Dad arrived on Corsica—to the *Gothic Line*, in Tuscany, in September.

P-47 Thunderbolts flying from Corsica were used as ground-attack aircraft in an air interdiction campaign against railroads and highways used to supply the Germans behind their lines. The Thunderbolts also acted as fighter escorts for raids by B-26 Marauders and B-25 Mitchell bombers. (The Mitchell bomber, by the way, was named for controversial early air-power visionary General Billy Mitchell, *who grew up in West Allis*—regarded by many as the father of the US Air Force. In 1924, Mitchell famously predicted war with the Japanese was inevitable, and would begin with a surprise aerial attack on Ford's Island at Pearl Harbor, Hawaii at 7:30 on a *Sunday morning*. He was off by twenty-five minutes, as the attack seventeen years later—on *Sunday*, December 7, 1941—occurred at 7:55 a.m.).

After fierce fighting at Anzio beach and at Monte Cassino, the Allies liberated Rome on June 4, 1944, a few days after Dad arrived on Corsica, and began an attack on the Germans' next line of defense.

Two days after Rome was liberated, the Allies launched the Normandy Invasion in northern France, opening a new front in the war. The Normandy Invasion was supposed to be coordinated with an invasion of Southern France from the Mediterranean called *Operation Anvil*; but the brutal road to Rome had taken too long, and *Anvil*, renamed *Dragoon*, was put off to August 15, 1944. Unfortunately, the invasion of southern France took away from the resources of the Allied armies in Italy as VI Corps—over one hundred thousand men—was switched from the Fifth Army in Italy to the Seventh Army for the invasion of southern France. To the surprise of the Germans, the Italian campaign ground to a near halt at the Arno River in Tuscany in September. The Germans in retreat

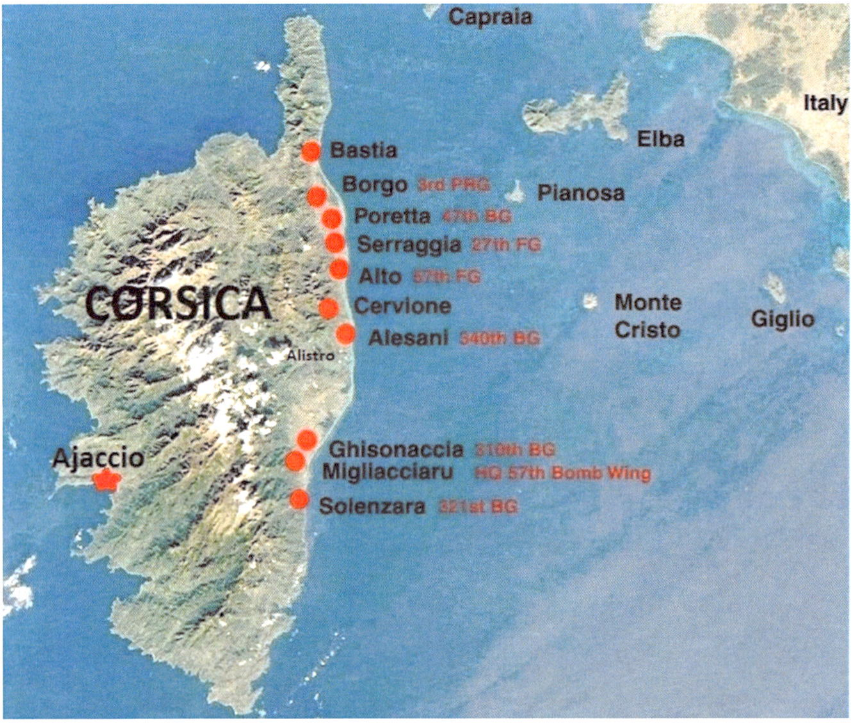

Allied airfields on Corsica. Dad's unit, Battery D of the 454th Anti-Aircraft Artillery AW Battalion, guarded Alesani Airfield of the 340th Bombardment Group near a little village called Alistro.

had destroyed all the bridges on the Arno except for the historic *Ponte Vecchio* in Florence.

Then the Germans retreated about twenty miles north to their Gothic Line. Fifth Army's shortage of manpower and equipment not only halted the Allied advance, but it also had major consequences for Dad.

Big things were also happening on the Eastern Front in 1944. Two weeks after the Normandy invasion, the Russians launched what was to be their largest offensive of the War, *Operation Bagration*, which broke the back of the German presence in the East, thanks in part to Hitler's foolish orders, overriding his generals, to try to hold a vulnerable salient, and in part to the Soviet Army's new mobility after being supplied with 150,000 trucks built in South Bend, Indiana by Studebaker Corporation, 50,000 Willys and Ford jeeps, and 26,000 Milwaukee-built Harley-Davidson motorcycles, mostly delivered through Allied-occupied Iran by way of the so-called *Persian Corridor*.

I wish I could remember the details, but I recall Dad telling me he once made a quick trip to **Egypt**. I think it was while he was around the flyboys from the bombers on Corsica that he was invited to see the pyramids. I suspect this happened while he was on Corsica, because that was where he was closest to the pilots. In any event, at some point while overseas he did get to fly to Cairo or Alexandria for a few days on a lark, probably the first plane ride of his life. I have no pictures from his trip to Egypt, or his time on Corsica.

Meanwhile, back home in the US, the Democrats held their nominating convention in Chicago in July 1944. Wartime leader Franklin D. Roosevelt—secretly diagnosed earlier in the month with congestive heart failure, coronary artery disease, and severe hypertension—had little opposition for nomination to an unprecedented *fourth term*. Party conservatives, however, viewed popular, but controversial two-term Vice-President Henry A. Wallace of Iowa as too anti-segregationist, too pro-labor, too pro-Soviet, and too much of a religious kook for flirting with eastern mysticism—and he was dumped from the ticket in favor of Senator Harry S. Truman of Missouri, considered more of a known quantity. Truman, who had been an artillery officer in World War I, won the vice-presidential nomination on the second ballot (with FDR's approval and after many delegates were persuaded by party leaders to shift their votes). The Republicans had previously nominated Governor Thomas E. Dewey of New York.

There was an "old-enough-to-fight/old-enough-to-vote" movement during the War, but the 26th Amendment, lowering the voting age from twenty-one to eighteen, was not adopted until 1971. Dad, at only twenty years of age, was still too young to vote in 1944.

. . .

While Dad was on Corsica, the Germans were aware of troop and ship movements suggesting an invasion in August 1944. But they anticipated a landing at heavily defended Genoa, Italy's largest port, close to the industrial heartland of Milan and Turin.

Instead, the Allies and Free French troops landed between Cannes and Toulon on the Riviera in Southern France. Allied war planners had concluded the ports in Northern France and Belgium were inadequate

Google Earth view of where the Alesani Airfield was located, now just fields, and Alistro, where Dad's anti-aircraft battery was stationed in eastern Corsica. (©2017 *Google*).

to support the number of Allied troops necessary to win on the Western Front, so the real objectives of *Operation Dragoon* (formerly *Anvil*) were the French Mediterranean port of Toulon and the major seaport and railhead of Marseille at the south end of the Rhône River Valley.

Operation Dragoon, beginning in mid-August, was considered a major success—even though the British had opposed such an invasion because it took so many Allied resources away from Italy and frustrated Allied efforts to conquer Italy by winter. Winston Churchill had favored putting more effort into entering the Po Valley in Italy, landing a force in the Balkans, and trying to beat the Russians to Vienna; but the Soviets had backed the American plan of attacking southern France, which would leave more of Eastern Europe within the Soviet sphere after the War. Army Group 6, which undertook Operation Dragoon in Southern France, eventually joined General Eisenhower's two Army groups (12th and 21st), giving him *three* Army groups to attack Germany from the west.

As time went on, and the Allies more and more controlled the skies, the Allied airfields on Corsica needed less and less defending by anti-aircraft artillery units like Dad's. In fact, the War Department began thinking it had more personnel in anti-aircraft artillery than it needed.

And Fifth Army in Italy now was desperately short of troops.

On August 27, 1944, about three months after Dad arrived on Corsica, his battalion was relieved and turned over its Bofors guns and equipment to another anti-aircraft battalion. The next day, Dad was part of a motor convoy to the port of Ajaccio on the western side of the island. That evening,

Dad's battery boarded an old Italian liner converted to a troop transport called the *Argentina*—not to be confused with the American ship of the same name—and left port the following morning. After sailing through the Strait of Bonifacio, between Corsica and the Italian island of Sardinia, the ship stopped at La Maddalena Island, Sardinia, and finally arrived at the port of Naples, Italy on August 31, 1944. There, Dad's battery joined the other batteries in the battalion that had come on different ships.

Dad was in the vicinity of Naples for the next six weeks with time on his hands—while his battalion was undergoing an amazing transformation.

7

ITALY

On September 25, 1944 an extraordinary thing happened. Dad's battalion, the 454th Anti-Aircraft Artillery Automatic Weapons Battalion, was disbanded and reconstituted as the 1125th Armored Field Artillery Battalion. Dad was assigned to Battery A. The new unit was to be equipped with an entirely different weapon-system: the M7 armored 105mm self-propelled howitzer. The personnel, going from anti-aircraft to field artillery, would have to undergo extensive in-theater training. A lot needed to be done to make the new 1125th combat-ready.

The changeover from anti-aircraft to armored field artillery involved not only new equipment but personnel changes, as well, including new officers with training in field artillery. The new battalion commander was Lieutenant Colonel Maurice P. Shaver. The commander of Dad's battery was Captain John W. Edge. The battery also had two first-lieutenants, Oscar Peters and George Garlock, and a second-lieutenant, James Montgomery, Jr.

I have obtained a copy of a 2006 self-published book called *Metamorphosis in the Mediterranean* by a guy named Leonard Warner. Like Dad, he also was in the 1125th Armored Field Artillery Battalion; but he was in Headquarters Battery—unlike Dad who was in Battery A. According to Mr. Warner, the 1125th bivouacked near the Naples harbor where there were a lot of brothels, streetwalkers, dance halls, drinking and gambling establishments, and other amusements to separate the soldiers and sailors from their money.

As Dad was to learn shortly when he was transferred from Naples in the South to Tuscany, in the North, Italy was historically, culturally, economically, and politically almost like two separate countries, north and south. Called the problem of the *Mezzogiorno*, southern Italy traditionally was poorer, had less industry, less literacy, was more peasant-dominant under the *padrone* system, was more Catholic, more in favor of the monarchy, and more corrupt (under the *Mafia* in Sicily and *Camorra* in Naples, for example). Northern Italians viewed Southern Italians as less self-reliant and hard-working than in the North, which was more middle class and more like the rest of Europe. Southerners viewed Northern Italians as materialistic snobs without traditions.

Dad probably saw more poverty and more theft, corruption, and organized criminal activity in Naples than he experienced anywhere else in Italy. The war had hit the infrastructure and economy of Naples hard. Everywhere, children in rags, the *scugnizzi*, were begging, stealing, picking up cigarette butts discarded by GIs, and pimping for their sisters or female cousins who had turned to prostitution to put food on the table. Civilian survival depended on the thriving black market.

I think it was in Naples that Dad was "rolled." He told me he was in a dark alley one night and got hit over the head. When he *came to*, his

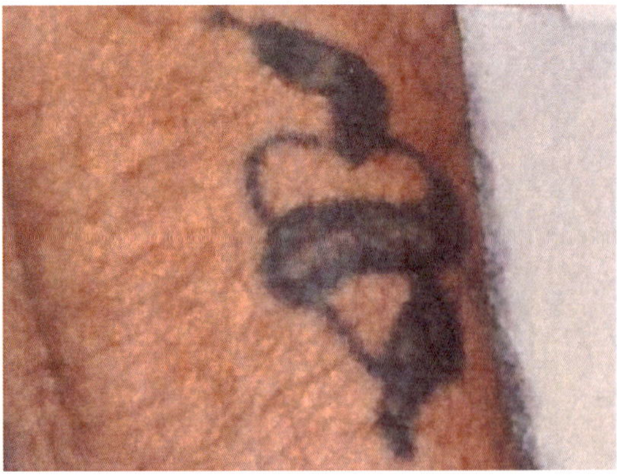

Dad's tat.

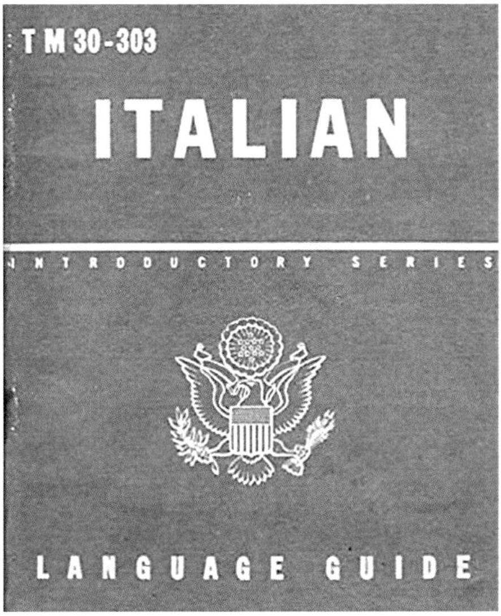

War Department training manual.

money was gone. I do not know if drinking was involved, but suspect it was. I am not sure, but it might have been in Naples that Dad got his tattoo. A rite of passage for a GI in those days, the tattoo on Dad's right forearm was of a snake wrapped around a heart, but I cannot recall his telling me the city where he had it done. Dad was only twenty at the time, not old enough to vote or legally drink liquor back home.

While in Italy, Dad tried to learn Italian, starting with a training manual.

Mr. Warner's memoir makes clear the soldiers overseas faced a lot of sexual temptations. Dad was generally silent on this subject, and I never asked him directly, probably not wanting to know the answer, but I got the impression—though he never directly told me so—that he had a sexual experience or two while in the Army. Like all new recruits, Dad was given a formal morality sex lecture when he first entered the service. He was also shown a film and given a pamphlet about prevention, recognition, and treatment of sexually transmitted diseases. Most likely, he was given follow-up lectures, formally or informally, at every new duty station. The Army was always concerned about soldiers getting venereal diseases, and the GIs frequently had to submit to notorious "short-arm" inspections. Although abstention was promoted, the Army had a more realistic attitude about the sexual behavior of young men in World War II than it had in World War I. Condoms were readily available in most command headquarters—no questions asked—or from medics. PXs carried condoms and prophylaxis kits, which took advantage of advances in antibiotics between the World Wars.

At the time Dad was in Naples, the city was wracked with typhoid, smallpox, and malaria. Dad told me in North Africa he had to take a drug called *Atabrine*, which turned his skin *yellow*, to guard against malaria. His immunization records show he was vaccinated against smallpox, as well as typhoid and tetanus, among other diseases. However, the records produced do not show any medical treatment. I do not have Dad's medical records, which may not even exist.

. . .

I remember Dad telling me about something of a supposed miracle he witnessed during his time in Naples. Every year, on the feast day of the city's patron saint, St. Januarius (*San Gennaro*, in Italian), a vial or ampoule of the saint's dried blood in a hand-held reliquary supposedly *liquefied* and was shown to the public in a ceremony at the Naples cathedral. Dad said he went to the cathedral when the liquid blood was displayed by the archbishop on the feast day of St. Januarius, (September 19)—but he was too far away to really see.

On October 18, 1944, Dad's battery, presumably with their M7 self-propelled howitzers, boarded LST 603 in Naples harbor and headed north to Leghorn—*Livorno* in Italian—in Tuscany, closest port to Florence. The flat-bottomed LST, or "Landing Ship Tank," was designed with two huge doors in the bow that opened up so tanks or other vehicles could be unloaded right onto a beach, if necessary. Over a thousand LSTs were built during the War, mostly in inland waterways in America's heartland. The LST Dad was on, for example, was made by the Chicago Bridge & Iron Company in Seneca, Illinois, on the Illinois River.

On the boat ride north, Dad told me he saw the Anzio and Nettuno beaches where Fifth Army Allied soldiers earlier in 1944 had made an amphibious landing in an effort to outflank the Germans who were blocking the way to Rome—only to be pinned down for many months with horrendous casualties caused by Luftwaffe sorties, enemy field artillery, and two super-heavy German K5 railway guns, beyond the range

An LST at Naples harbor during World War II. (*National Archives*).

of Allied artillery and hidden in a tunnel when not in use, so as to be invulnerable to Allied air power. The big guns—captured by the Allies after the Anzio breakout—were called "Anzio Annie" and "Anzio Express" by the GIs and could fire a 562 lb. shell nearly 40 miles.

When the Germans abandoned the port of Leghorn in Tuscany in late July of 1944, in front of advancing Fifth Army troops, they sabotaged it even worse than they had the port of Naples with mines, booby traps, sunken ships and debris, destruction of cranes, depots, and rail connections, etc. The task of clearing the port and making it operational fell to what was called the *Peninsular Base Section* of the US Army Corps of Engineers. The port at Leghorn had only become operational less than a month before Dad's LST arrived.

Dad, and the other front-line troops, probably never knew how much they were dependent on the Peninsular Base Section, a combination of military (including some former "ack-ackers," like Dad), and US civilian and local Italian engineers and laborers operating in the rear who were responsible not only for clearing and operating port facilities, but also distributing petroleum products, building pipelines for gasoline and water, repairing and building roads and bridges necessary for moving supplies, constructing hospitals, repair shops, rest camps, and prisoner-of-war cages, operating sawmills and quarries to provide building materials, firefighting, map-making, and even securing leases with local Italians for properties needed by Fifth Army and its units.

8

THE M7

Though American-made, the new weapon operated by Dad's battalion, the M7, first saw combat in Egypt—*with the British.*

Winston Churchill happened to be meeting with President Roosevelt at the White House on June 21, 1942, when he heard the news about the Brits' worst disaster in North Africa: the fall of the Libyan port of *Tobruk,* near the Egyptian border. Churchill did not hide his disappointment from President Roosevelt. The defeat involved the surrender of 33,000 Allied soldiers to perhaps half as many Axis troops, and Churchill deemed the defeat "a disgrace." It meant the Nazi *Afrika Korps* would now have a shorter supply line to attack the British Mediterranean port and air base at Alexandria, Egypt, defending Cairo and the Suez Canal. If Suez fell, Britain would be deprived of a shortcut to its vital resources—and give the enemy access to Mideast oil.

President Roosevelt asked what the US could do to help, and Churchill requested *tanks.* Army Chief of Staff General George Marshall was called in and offered to send the Brits 300 Sherman tanks *plus 100 M7s*—the Army's new armored self-propelled howitzer. With the Mediterranean still too dangerous, the tanks and M7s reached Egypt around the Cape of Good Hope via the Indian Ocean and Red Sea, in time to contribute significantly to a British victory at the critical Second Battle of El Alamein, Egypt in Autumn of 1942—the turning point in the North African campaign.

. . .

The operator of the .50-caliber machine gun in the distinctive, round, armpit-high, armored area on the right side of the M7 reminded one Brit in Egypt of a priest in a pulpit, and the M7 forever-after was called "The Priest." It was considered faster, more maneuverable and with greater range than a clumsy, British-designed, high-profile self-propelled howitzer called "The Bishop." In North Africa, the M7 proved effective in helping the British knock out the famed, dual-purpose (anti-aircraft/anti-armor) German 88 Flak guns, considered the best anti-tank guns of the war.

Originally designed for the Lee tank carriage, the M7 Priest by 1944 was built on the chassis of a Sherman tank; but technically it was not a tank in that it was completely open on top, not heavily armored, and its armament consisted of the .50-caliber machine gun and an M2A1 105mm *howitzer*, instead of a *gun* like a tank.

I learned something about artillery when I was stationed at Fort Sill, Oklahoma back in 1969. Without question, *guns* and *howitzers* have some overlap, but the fundamental difference is that the gun usually fires with a very high muzzle velocity directly at the target, quickly overheating the

The M7 Priest armored self-propelled 105mm howitzer. (*National Archives*).[c]

The M7 Priest from above. (*National Archives*).[c]

barrel and limiting the rate of fire. The howitzer, on the other hand, more or less fires *indirectly*, lobbing the shell at the target with the minimal amount of muzzle velocity to get it there and can be fired repeatedly with less risk of overheating.

Guns ordinarily fire *directly* at a target visible from the gun's position—often targets that are moving, like a tank or ship or airplane. Howitzers, by contrast, usually fire *indirectly*, assisted by forward observers who radio target information to a fire-control center, which gives directions to the howitzer batteries—out of sight of the target. In Dad's time, like today, American infantry and armor units did not like to operate outside of the range of friendly artillery, available at a moment's notice in any weather.

The M7 Priest, of which over four thousand were produced, had a top speed of about 24 mph on the road and 15 mph off-road. Its maximum range for firing a 33 lb. high-explosive shell was over *seven miles* (12km), which meant Dad's unit would not be far behind the front-line infantry or armor they would be supporting, often within range of the German artillery, which always had as an objective knocking out Allied artillery batteries, like Dad's.

An armored field artillery **Firing Battery**, like Dad's, headed by a captain, consisted of **six** M7 Priest self-propelled 105mm howitzers and around one hundred **men**. *That is about how many men Dad was responsible for feeding.* Depending on the mission, the battalion's fire-direction control could coordinate all three firing batteries in the battalion—or the firing batteries, like Dad's, which also had their own integral fire-direction control sections, could act *independently*, if needed.

The hundred, or so, soldiers in Dad's battery were broken down into twelve "sections." Each of the six M7 Priest howitzers was a section of seven men and was assisted by a Fire-Direction Control Section and an Ammunition Section. Scouting was done by a Reconnaissance Section. Battery Headquarters was made up of a Headquarters Section, a Maintenance Section, and the **Administrative, Mess, and Supply Section**—the section Dad was in—which consisted of thirteen enlisted men including the supply sergeant, and Dad, the mess sergeant (as well as his two cooks and cook's helper).

. . .

Dad's **Armored Field Artillery Battalion**, headed by a lieutenant colonel, had a total of *five* batteries: three *firing batteries* (like Dad's, which could operate independently)—a total of 18 M7's—plus a *Headquarters Battery*, core of the operation, which included headquarters, communications, a battalion fire direction center, and two observation aircraft—and finally a *Service Battery*, with its numerous trucks, half-tracks, support vehicles, drivers, mechanics, etc., responsible for transporting all the necessary ammunition, fuel, and food and water for all five batteries in the battalion and maintaining all the vehicles. The battalion also had a small *Medical Detachment* with a doctor.

In all, the field artillery battalion, consisting of about five hundred **men**, was the basic fighting unit in artillery, and was attached to infantry and armor, which were organized into *much larger* types of units. These included **Regiments** (three or more infantry or armor battalions) and **Divisions** (three or more regiments). Two or more divisions could be organized into a **Corps**.

Two or more corps became an **Army**, and two or more armies became an **Army Group**, the largest type of fighting unit in a **Theater**. The Mediterranean Theater had only one army group in Italy, the 15th, which consisted of the British-led Eighth Army and the US-led Fifth Army.

9

STAFFOLI

After Dad got to Tuscany, he spent six weeks, or so, in the fall of 1944, in what became one of his favorite places in Italy.

The Arno River flowed west from Florence through Pisa to the sea. The Germans had retreated north of the Arno River to their *Gothic Line,* a series of fortifications which took advantage of the mountainous terrain of the Northern Apennines. The Allied soldiers were all aware the objective was to break through the Gothic Line and proceed north to the other side of the Apennines, the Po Valley—which was really more like a fertile *plain* between the Apennines and the Alps—about sixty miles wide and four hundred miles long from west to east. North of that were the Alps, where the Germans could dig in again.

After the tough mountain-fighting the Allies had encountered earlier in 1944, Dad and the other soldiers in his unit, who had never experienced such combat, must have been wary of the danger and hardships that lay ahead.

In the fall of 1944, newly-arrived Fifth Army units, like the 1125th, were scattered throughout Tuscany to undergo training. Dad's battery was stationed near an out-of the-way little Tuscan town called *Staffoli* on a

Dad's sea journeys in the Mediterranean in 1944, from Oran, Algeria, to Ajaccio, Corsica, to Naples, Italy, and north to Leghorn (*Livorno*) in Tuscany.

provincial road north of the Arno River, in the foothills of the Apennines, west of Florence (birthplace of the Renaissance) and east of Pisa (home of the Leaning Tower). The people of the town were friendly to Americans, and Dad did his best to try to communicate. As a mess sergeant, he was in a position to barter with the locals for whatever fresh food was available, and he recognized speaking a little of the language would help.

Dad brought home this photo of street kids in Italy.[c]

After years of food rationing under the Fascists, many of the local people in Staffoli were hungry, and Dad saw to it they had something to eat. Especially hard-hit everywhere were the children; Dad, like the other GIs, felt sorry for them and tried to

Dad in Staffoli. *Fall 1944.*

Dad in front of S. Michele Arcangelo (St. Michael Archangel) church. Staffoli (near Pisa), Italy. March, 1994.

help. A "street kid" himself, Dad, I am sure, was particularly sympathetic to the plight of the children.

One Italian man Dad befriended was named Bruno Bellagamba. Dad gave him table scraps, leftovers, and extra food; and he credited Dad with saving his family from starvation.

Fifty years later, in March of 1994, my wife Cheryl and I accompanied Mom and Dad to Staffoli where we had a magical day. We first went to the village church, St. Michael Archangel, and created such a sensation as

Dad and Mom in March, 1994 with Bruno Bellagamba's widow and other friends in Staffoli, Italy.

foreigners that the pastor was called. Eventually, we mentioned Dad was a World War II vet who wondered what happened to a friend of his named Bruno Bellagamba. Phone calls were made, and we ended up meeting Bruno Bellagamba's widow and family, going to their house, to the cemetery to see Mr. Bellagamba's grave, to a site in one of Dad's photos, and to visit another relative.

My college-Italian was rusty, but much better than Dad's. From what we were told, it was clear the Germans had occupied Staffoli for a time before the Americans came. Fifty years is a long time, and Dad was no longer the twenty-year old *sergente* he had been in 1944. Not that it mattered, but I was not sure the friends we met even recognized Dad. It was, however, for Dad, a real nostalgia trip.

From what I have learned since—the 913th Field Artillery Battalion was the first American unit to operate around Staffoli in early September 1944, after the Germans retreated from the Arno River north to

In the field. Dad in battle gear with his carbine. The truck behind him is likely his kitchen truck. To the left is a tent that may have been used for mess purposes. Photo possibly from forest land around Staffoli in November, or so, 1944.[c]

This is one of the strangest pictures Dad brought back. I believe it shows the back end of his kitchen truck parked in some rural area (note hay) outside a sheltered garage or stable area. In the modified 2 1/2 ton 6 x 6, the hoops for the canvas top were extended so men could stand up inside the truck; and the normal inside seats along the bed were moved outside and used as shelves for food preparation or for eating standing up. The back door has "Keep Out" written above it. Note the 5-gallon *Jerry can/Blitz* can on the back, and others in the lower right of the picture. The bench in the picture could be used while eating. Since the six M7s in Dad's battery were manned 24/7 by enough personnel to perform a fire mission, eating in the field was done in shifts.

their defenses at the Gothic Line in the Northern Apennine Mountains. Staffoli was also briefly a headquarters for the Allied IV Corps and also served for a short time as a base for the Brazilian Expeditionary Force.

So, the people of Staffoli saw a lot of foreign soldiers from 1944–45.

Dad was probably not aware of it, but while he was at Staffoli in fall of 1944, with the Allies needing less anti-aircraft artillery, Camp Davis in North Carolina, where he had spent most of the previous year, was already in the process of being dismantled.

In the rush through basic training back at Camp Davis, Dad had difficulty qualifying with the M1 Garand rifle, standard-issue for US *infantry* troops—finally making "marksman" (the lowest qualifying badge). I remember him telling me how he much preferred the smaller, lighter .30-caliber M1 *carbine* (for which he qualified as "sharpshooter," the rank below "expert"). I don't know that he ever had to use his carbine,

M-1937 Field Range.

which was standard-issue in Dad's *artillery* battalion.

Dad was in charge of two cooks and a cook's helper, plus whatever enlisted men were needed for KP (kitchen police) duty. He was assigned a two-and-a-half-ton 6x6 kitchen truck, which contained the equipment necessary to cook meals. Often, when on the move, the kitchen truck would tow a trailer with extra food or other supplies, or a water tank with potable water. Food from the nearest quartermaster dump was delivered to Dad by trucks from the Supply and Transportation Platoon of the battalion's Service Battery.

One of the things Dad's unit had to learn while doing field exercises in Staffoli was the art of camouflage. The whole idea with the howitzers was to keep them hidden so the enemy did not know where the Allied artillery fire was coming from, and therefore could not fire back. This meant not only the M7 Priest howitzers had to be hidden, but also supporting equipment like the fire-direction centers, usually on half-tracks,

Thanksgiving outdoors with an M-1937 field range. (*U.S. Army Quartermaster Museum*).[c]

Jerry cans (Blitz cans): for water (left) and gasoline (right).

Cooks in a kitchen truck with two field ranges.[c]

and other supporting vehicles, like Dad's kitchen truck. Before Dad's battery would move, scouts would go ahead and find a place to locate.

The heart of any kitchen truck was the versatile M-1937 field range. Most kitchen trucks had two of them. It came with stock pots, and a large roasting pan with cover. The range could be used for baking, roasting, and—with the roasting pan cover placed on top—as a griddle. The fuel was gasoline, put under pressure with air from a hand pump—not unlike today's Coleman camping stove. Either leaded gasoline or unleaded white gas (Coleman fuel) could be used—though leaded gasoline required special filtering. If necessary, the range could even use wood for fuel.

The gasoline to operate the field range was transported in a 5-gallon (20 liter) can, called a *Jerry can* or *Blitz can* because it was originally a German invention. The Allies reverse-engineered it, and produced millions and millions of Jerry cans. The Jerry can came in two versions, one for gasoline and one for water. Dad would have used both kinds. Gasoline

US Army canteen and cup with folding handle. The canteen fit into the cup. With the cup handle folded, the canteen and cup fit into a belt pouch.

The mess kit, too, had a folding handle. The cover could be placed inside the main part to separate the food into two areas. The cover and utensils fit over the handle and the whole kit could be dipped into hot water for cleaning.

Immersion heater with garbage can.

was moved by truck and rail. By March 5, 1945, the Corps of Engineers actually constructed a gasoline *pipeline* from the port at Leghorn to Pistoia at the foot of the Apennines.

In the field, each soldier carried a canteen, with a cup, and a mess kit.

The heart of cleaning up after eating was the immersion heater, again gasoline-fueled. The heaters were placed in a series of garbage cans. The first can (without heater) was for scraping off what food particles could be removed. Next came a can, usually with warm water, that contained detergent. Next came a warm water rinse, and then a boiling hot rinse.

Often, a garbage can full of boiling water would be available before eating, too, so the soldiers could sterilize their mess kits before they ate. In the field, garbage cans were usually emptied into a "slop pit," a large hole in the ground that was covered over with dirt when full.

A group of GIs using immersion water heaters to clean their mess kits. (*WW2 U.S. Medical Research Centre*).[c]

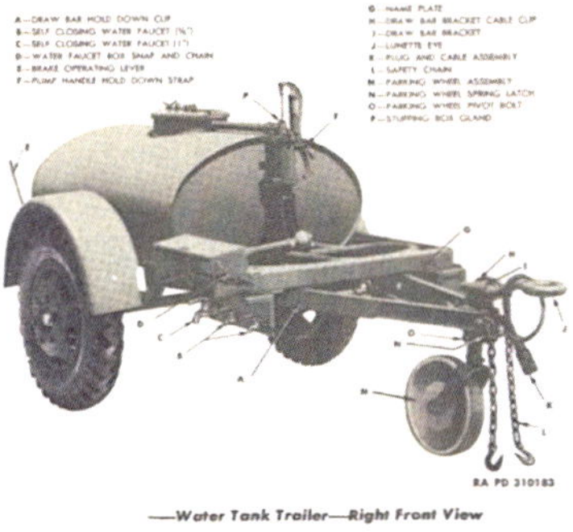

The 250-gallon water-buffalo trailer.[c]

Some of the gasoline-operated equipment, like the immersion heaters, had problems, especially when using leaded gas. I remember Dad telling me at one time he invented some kind of a burner to heat the water from below. I wish I could remember more of what he said, but it would be like Dad to figure out a better way of doing things, or at least a way to get by.

Water, of course, was absolutely vital in running a field kitchen and would usually be transported in a 250-gallon water trailer, sometimes called a *water-buffalo*. (Major manufacturer of the G-527 water-buffalo trailer was the Ben-Hur Mfg. Co. of Milwaukee). The Corps of Engineers set up portable water purification trucks at streams, wells, or other water sources, which were called "water points," where the 250-gallon water-buffalo and 5-gallon water Blitz cans could be refilled with potable water. Pipelines were even built from water wells to provide water for filling Blitz cans and water-buffaloes. From a water point, the water-buffalo trailer would be towed to wherever Dad's mess was set up.

Dad had to set up his mess under many different conditions. One of the

Testing shoepacs. As with most of Dad's photos, this one is undated and does not identify the muddy area where it was taken.[c]

lieutenants would have been designated the mess officer, and Dad would have been responsible to him. The officers had their own mess whenever possible, and I am sure Dad provided special treatment. However, in the field, the officers ate the same as the enlisted men. The tradition was for the enlisted men to eat first, and the officers last. Though not considered a senior non-commissioned officer (NCO), Dad, as a staff sergeant, did favors for the other NCOs, like the first sergeant, who was responsible for leaves and passes, and the supply sergeant who got the latest in uniforms and equipment.

I think while Dad was in Staffoli he tried his best to get fresh food for the hundred or so men he had to feed. It is hard to say what was available, however. When the Germans left, a little over a month before Dad got there, they probably took whatever fresh food they could. Fifth Army also tried to supply fresh food, when possible.

Another undated photo with no location indicated, possibly from Staffoli. I do not know where Dad got the coverall or what appears to be an Italian dueling sword.

The Americans had a generous attitude about food. According to Undersecretary of War Robert P. Patterson, only about two-thirds of the food the US Army purchased actually went to our troops. The rest went to feeding the troops of some of our allies, like the Free French, and prisoners of war, of which there were many. We also, in many cases, had to feed people in lands we liberated—at least until they could get their own economy going again. If Dad was able, I am sure he bartered canned goods, flour, salt, cigarettes, coffee, chocolate—whatever he could—to get fresh eggs, chicken, cheese, fruit, vegetables, baked goods or whatever else the locals in Staffoli were willing to trade. And if he could help the hungry Italian people, he did so.

With winter approaching, Fifth Army distributed woolen underwear, waterproof winter clothing, and "shoepacs"—boots with a rubber bottom and waterproof leather top. During November, Tuscany was hit with a number of heavy rainstorms, and snow began to appear in the higher elevations of the Apennines.

With winter setting in and amid rumors Fifth Army's big push through the Northern Apennines would not come until spring, Dad's sojourn in Staffoli was unexpectedly about to end. A war was going on, after all. The in-theater training had gone well; the conversion from anti-aircraft artillery to field artillery was complete, and the higher-ups now had plans for Dad's unit—Battery A, 1125th Armored Field Artillery Battalion—that called for months of hard fighting ahead.

10

CAPUGNANO

Howitzer field artillery came in three basic designations: *heavy*, which was 8-inch, or larger; *medium*, which was 155mm; and *light*, which was 105mm—like in Dad's unit—or less.

Dad's battalion of 105mm self-propelled howitzers was what was considered a Fifth Army "bastard unit," an independent or *non-divisional* battalion, assigned by Army HQ to *IV Corps*. Generally, infantry and armor divisions had their *own* artillery. Divisions had one 155mm artillery battalion and three 105mm artillery battalions—towed for infantry, self-propelled for armor—supporting the division's three regiments and under control of the division artillery commander.

An independent battalion, like Dad's, however, could be attached by Army (or Corps) headquarters to any division, regiment, or smaller unit that needed additional fire-power, for whatever reason.

A previous independent M7 battalion (69th Armored) went from Italy to France with VI Corps in August 1944, before Dad's battalion was created.

As nearly as I can tell, while Dad was in Italy, his battalion was *unique*—**the *only* "non-divisional" armored field artillery self-propelled 105mm**

howitzer battalion *in the whole Fifth Army.* This unusual status would give Dad a unique perspective on the War in Italy.

On December 7, 1944, Dad's battalion was ordered to move from Staffoli about fifty miles north on various state highways and regional and provincial roads to a small mountain village called Capugnano just outside a larger spa town called Porretta Terme. Porretta Terme, which had maybe three to four thousand residents, was the headquarters of the Brazilian Expeditionary Force, and Dad's battalion was ordered attached to the 1st Brazilian Division.

You may be wondering: What were Brazilians doing in Italy?

Who were the *Allies*, anyway? And whom were they fighting?

As early as 1936, Benito Mussolini, Italy's Fascist dictator, proclaimed a *Rome-Berlin "Axis"* existed, around which the rest of Europe would revolve. In 1939, Tokyo, which also had expansionist military ambitions, joined the Axis, though it played almost no role in the Mediterranean.

After the Allied success over the Axis powers in North Africa, Sicily was attacked in July of 1943 as the first step in invading Italy. British prime minister Winston Churchill called Italy the "soft underbelly" of the Axis and correctly believed an invasion would result in Germany committing resources and manpower to Italy that otherwise would have been used against the Russians on the Eastern Front or against the English, Americans, Canadians, Free French, and others preparing the invasion of Northern France at Normandy.

Benito Mussolini, a former teacher and prominent Socialist newspaper editor, whose mother had been a teacher, broke with the Socialist party by backing Italy's entry in World War I. Rejecting the class warfare of orthodox Socialism, he favored ultra-nationalism and founded the National Fascist Party in 1921. When parliament was unable to form a government, King Victor Emmanuel III—fearful of the Socialists, Bolsheviks, and anarchists—made Mussolini prime minister of Italy at age thirty-nine in 1922, even though the Fascists were only a minority party. Within a few years, Mussolini, meeting twice-weekly with the king (who had to sign all decrees), managed to consolidate power, outlaw all other parties, govern without parliament, and develop a cult of personality—used as a model by Germany's Adolf Hitler, a vegetarian, nonsmoking teetotaler, who was head of the Nazi Party.

Hitler's father had changed the family name from Schicklgruber to Hitler before Adolf was born. (Imagine: *Heil, Schicklgruber*!). Born and raised in Austria, Adolf Hitler, who had a middle-class upbringing, mostly around Linz, did not move to Germany (Munich, Bavaria) until 1913, when he was twenty-four *years old*—after both of his parents had died and the Austrian army drafted him.

Although a successful student early in life, and a life-long reader, Hitler as a teen despised authority, was considered lazy and uncooperative at school, could not tolerate being criticized—and had a violent temper. He flunked a year of high school and eventually dropped out with failing grades at age sixteen. After a year of idleness and with his widowed mother diagnosed with terminal breast cancer, he took money inherited from his late father and moved to Vienna in hopes of getting admitted to the Vienna Academy of Fine Arts. Better at sketching than painting, he dreamed of becoming a famous artist and was crushed when the Academy rejected him (twice) as not talented enough. His devout Catholic mother had given birth to six children, but only Adolf and his younger sister Paula survived to adulthood. He also had a half-brother, Alois, and sister, Angela, from his father's prior marriage.

The half-brother, Alois, and his Irish first-wife, had a son who was raised in England, but also spent several years in Germany, meeting his uncle Adolf a few times. On November 4, 1939, the nephew, William Patrick Hitler, who had soured on Nazism, addressed the Wisconsin state teachers' convention in Milwaukee and estimated eight out of ten Germans were secretly opposed to his uncle. The young man was particularly upset by the scandal involving his cousin, Hitler's niece, Geli Raubal, daughter of Hitler's half-sister, Angela. On September 18, 1931, Geli, twenty-three, who lived with Hitler, was found dead in Hitler's Munich apartment, shot with Hitler's pistol after an altercation with her uncle. Officially a *suicide,* the death and Hitler's role in it have fueled speculation to this day—and threatened Hitler's political future at the time.

For several years after his mother's death in 1907, Hitler, who never had much of a regular job, lived aimlessly in Vienna off his meager inheritance and by selling, when he could, small drawings and watercolors, mostly of urban buildings. Vienna was the capital of the *Austro-Hungarian Empire*—which included Hungary, Bohemia, Croatia, parts of Poland, Romania, and Italy—and what was to become Yugoslavia and Czechoslovakia. Austria's defeat in a war with the Kingdom of Prussia in 1866 caused the numerous German-speaking kingdoms, duchies, and other principalities north of the Main River to join the North German Confederation, dominated by Prussia. The next year, the weakened Austrian Habsburg monarchy, to placate rebellious Hungary, restored Hungary as a kingdom within the empire and created a dual-monarchy—the Habsburg emperor of Austria also becoming the king of Hungary. A few years later, Prussia's victory over France in the Franco-Prussian War caused South German principalities—including Bavaria, which fought for the Prussians—to also join the North German Confederation, an economic union, which in 1871 under Chancellor Otto Von Bismarck, became the *German Empire*, headed by Prussia's king, Kaiser Wilhelm I.

The unification of Germany isolated German-speaking Austrians, now a minority in the multi-ethnic Austro-Hungarian Empire.

The new German Empire consisted of four kingdoms, plus numerous duchies, grand duchies, and principalities—all headed by monarchs—and three city-states. It grew to be the largest and most industrialized country in Europe during the later nineteenth and early twentieth centuries; and many minority ethnic Germans in Austria-Hungary—who represented only about a quarter of their country's diverse population—longed to be part of Germany. Among these was Adolf Hitler.

An important influence on the development of Hitler's politics at this time was a wealthy Austrian politician named Georg Ritter von Schönerer, an admirer of Otto von Bismarck, the Prussian chancellor who had unified Germany. Von Schönerer was anti-Semitic, anti-Slav, anti-Catholic, anti-Habsburg, and anti-democracy. He founded the *Pan-German Society* to promote unification of all German territories in Austria with the German Empire. His followers called him *Führer* and greeted each other with "*Heil!*" Extremists like Schönerer were in the minority, however, as most ethnic-German Austrians still believed in Austrian sovereignty and were loyal to the Habsburg monarchy—unlike Hitler.

As Hitler's money dwindled during his years in Vienna, he could be found on the street *begging*—unshaven, with long hair, tattered clothes, and holes in his shoes. He slept on park benches, in flophouses, at a men's hostel, or in a homeless shelter. He later called his five years in Vienna "the saddest time in my life," but said the experience convinced him Marxism and Judaism were the two greatest dangers to the German people.

On his arrival in Vienna, he at first found employment in the building trades, but rejected a life of manual labor when he refused to join labor unions, which at the time were dominated by Socialists, whom he despised. Later in life, he credited the poverty and economic uncertainty of his Vienna years with giving him insight into human nature and into the people he would later represent and govern as a politician. When he turned twenty-four in 1913, and obtained the small final installment of his inheritance, he moved from Austria, where he was wanted for draft-dodging—sometimes living under assumed names—to Munich, Germany, in Bavaria. He rented a room and continued to live by selling his sketches and watercolors.

Hitler supposedly was reluctant to be conscripted into the Austrian Army where he might have to take orders from Slavs, Hungarians, or other non-German ethnic minorities that were part of the Austro-Hungarian Empire. Bavarian police arrested him and returned him to Austria. Charges of draft evasion were dropped, however, when he failed a physical examination. He then returned to Munich, where—unlike in Vienna—the people spoke a dialect of German similar to his—and put his Austrian life behind him for good.

Ironically, found *unfit* for the Austrian Army, he enlisted in the Royal Bavarian Army in Munich when World War I broke out in 1914. Still an Austrian citizen, he was accepted—some say by a clerical error—and took an oath to King Ludwig III, of Bavaria, as well as to the Kaiser. The Bavarian Army, subject to the *Imperial Army* in time of war, taught him discipline and gave him what he saw as a purpose in life.

At the end of October 1914, after a few weeks of inadequate training, and wearing only cotton oilcloth hats instead of helmets, Hitler's unit, the 1st Company of the 16th Bavarian Reserve Infantry Regiment, was thrown into battle in Belgium against the British—considered a fellow Germanic people—in what is now called the First Battle of Ypres. Somehow, Hitler, as a common infantry rifleman, survived the shelling and machine-gun fire, while two-thirds of his regiment were cut down in only five days of fighting, and the regimental commander killed. Hitler was quickly promoted from private to *Gefreiter*—approximately equivalent to

Private First Class (PFC), a rank *below* corporal—the only promotion he ever received.

On November 9, 1914, he was transferred from the 1st Company to Regimental Headquarters. He was assigned as an *orderly* or "gofer" for the officers. He mostly served as a *dispatch runner*, carrying messages, on foot or by bicycle, from Regimental Headquarters to the headquarters of the various battalions in the Regiment—though sometimes he also served as a *bodyguard* when officers went to the front. He remained an orderly/dispatch runner until the war ended four years later. Some say he refused any further promotion. After all, he had time for his sketches and watercolors and had a cushy job compared to front-line infantrymen in the trenches—but it was not without danger. By the end of the war, Hitler not only received a Wound Badge, but also Iron Crosses (Second and First Class) for courage. The Iron Cross *First Class 1914* was an unusually high award for a lower enlisted man like Hitler. He had been put in for it by his Jewish commanding officer, and Hitler was very proud of it. Affixed to his left breast pocket below his Nazi Golden Party Badge and above his black (3rd class) Wound Badge medal, it was the most prominent decoration he wore on his military uniform as Führer. By most accounts, he had been a good soldier during World War I, though his fellow soldiers always viewed him as a loner and a bit odd.

When the Armistice ending World War I took effect on November 11, 1918, Hitler's regiment was in Belgium; but Hitler, temporarily blind, was in a military hospital in Northern Germany recuperating from a British gas attack. A week and a half later, he went by train via Berlin back south to Munich, which at the time was in a state of *chaos*. The German Empire, the Kingdom of Bavaria, and the Royal Bavarian Army had by now ceased to exist.

Even before the Armistice, Kaiser Wilhelm II of the Hohenzollern dynasty had been forced to abdicate and flee to Holland by a civilian coup mostly disorganized Social Democrats supported by disgruntled military elements sick of war. The Allies, led by the US, had refused to negotiate an armistice with the Kaiser—insisting on what we today would call "regime change." So, signing the Armistice politically for Germany were representatives on behalf of a six-man *interim* civilian government made up of three moderate Social Democrats and three

Marxist radicals—who supposedly represented the whole defeated German Empire.

General elections set for January of 1919 were to elect representatives to a new national parliament. Former *monarchies* in Germany, like Bavaria—now only *states* in the new democracy—also set elections for early in 1919 for their own state parliaments. Though the Armistice had ended the fighting, it would be up to the newly-elected federal government in Germany to negotiate a treaty in Versailles, France to bring the war to a formal conclusion. Drafting and signing took over *seven months*, and the Treaty did not become effective until January 10, 1920.

Hitler was still hospitalized when Bavaria's aged King Ludwig III, of the Wittelsbach dynasty that had ruled Bavaria for over seven hundred years, fled to Austria during a local civilian/military revolt that had made Munich unsafe for the monarch. On November 12, 1918, from exile, the King declared all civilian and government officials and soldiers, like Hitler, released from their oath of loyalty to him. The new Socialist government in Munich considered the King's declaration an *abdication*. Without its king, the Royal Bavarian Army, in which Hitler was a soldier, no longer existed.

The military in Bavaria, as elsewhere, came under civilian control of the new national government. Pursuant to terms of the Armistice, Hitler, like millions of others, expected to be demobilized when he returned to Munich. However, having no family to go home to, and no prospects as a civilian, he asked if he could stay on as a soldier in the new army for a while, a request that eventually had dramatic consequences—for him and for the world.

About a month after his arrival in Munich, Hitler was assigned to guard-duty near the Austrian border at a Russian-French-British POW camp in the process of closing down. He was apparently there on January 19, 1919, when the new German republic, as an initial step in drafting a constitution in centrally located Weimar, held elections—for the first time giving German women the right to vote. Hitler, an Austrian citizen, was actually eligible to vote under a special reciprocity agreement between Germany and Austria. At the time, the German part of Austria was trying to be annexed to Germany, but the Allies, later in 1919, forbade such a unification.

In February 1919, nearing thirty, Hitler was sent back to the regimental barracks in Munich, where he lived comfortably. He expected to be discharged in May from the new federal army, called the provisional *Reichsheer*, still in the process of demobilization. Most of his days were spent inspecting gas masks, except when he was called out to guard the main Munich railroad station after a failed coup attempt.

The army, still dominated by officers from the nobility and naturally right-wing, had a political need to show the new democratic government, dominated by left-wing Social Democrats, it was *democratizing*—becoming more responsive to the ordinary recruit. To foster the trust soldiers must have in their officers for necessary discipline, every company was required to elect an enlisted man as a spokesperson, called a *Vertrauensmann*, to voice complaints to the officers and intercede with them on behalf of ordinary soldiers—the way a shop steward might represent workers in a factory. Hitler was elected *Vertrauensmann* for his company, which gave him special status with his peers and special access to the officers—in a sense, his first formal experience as a *politician*.

While the new Weimar central government was dominated by moderate Social Democrats, the former Kingdom—now State—of Bavaria, was run by more radical Social Democrats, and was called the *People's State of Bavaria*. The weak coalition government was headed by a Jewish former journalist with a long beard, Minister-President Kurt Eisner, who had engineered the ouster of Bavaria's King Ludwig III. However, Eisner, viewed as a radical by those on the right, and not radical enough by those on the far left, soon became unpopular and was assassinated by a right-wing nationalist February 21, 1919, on his way to resign—a month after his party suffered humiliating defeats in the January Bavarian and national elections. The same day, the head of the newly-elected more moderate Social Democrats was shot and critically wounded in the Bavarian Parliament chamber by left-wing radicals.

Eisner was succeeded as Minister-President by his education minister, moderate Social Democrat Johannes Hoffmann, who had removed the Bavarian education system from the supervision of the Catholic Church. Hoffmann also immediately came under attack from both the right and the left. The Bavarian government was at war with itself. As outbursts of violence, general strikes, and anarchy reigned in Munich,

Hoffman took the loyal elements of the government—Social Democrats and centrists—and fled one hundred forty miles north to Bamberg.

In early March of 1919, Vladimir Lenin hosted an international gathering at the Kremlin in Moscow called the 1st Congress of the Comintern—attended by representatives of more than two dozen countries—designed to spread Bolshevism throughout the world.

On April 6–7, 1919, inspired by Lenin's call for world revolution, and by a Communist revolution in Hungary, German Jewish dramatist Ernst Toller organized workers and engineered a Communist takeover in Munich. He proclaimed the establishment of the *Bavarian Soviet Republic*. Within a week, a more radical Jewish Communist, Eugen Leviné—actually born in Russia and with direct links to Lenin—took charge.

As Bavaria descended into civil war, the Bamberg government made several unsuccessful military attempts to oust the Soviet government in Munich—which had its own "Red Army" including some of Hitler's fellow soldiers from barracks in and around Munich. (Hitler was invited to join, but refused).

Fearful the Bavarian revolt might lead to a Soviet Bolshevik takeover of the whole country, and fearful that soldiers in Bavaria had been tainted by Bolshevism, the federal government in Berlin called on *Reichsheer* troops outside of Bavaria, as well as local right-wing armed militias called *Freikorps,* made up of *veteran* officers and enlisted men, to put down the uprising. Machine guns, flame-throwers, armored cars, even heavy artillery were used in the urban fighting. By May 3, when Munich was "liberated," more than six hundred had been killed, over half of them civilian "collateral damage." The fighting was followed by numerous executions.

Hitler, living in a Munich barracks, had no apparent direct connection to the fighting, but the political turmoil all around him could not help but reinforce his already-formed anti-Semitic and anti-Bolshevik views. To the man who was to become the most evil person of the twentieth century, the revolt served as a concrete example of what he perceived as a threat to the German people posed by a foreign—i.e., Soviet or Bolshevik—political and economic ideology, which he believed was dominated by Jews.

The new German army, which at first had tolerated political organization by the soldiers, now also viewed Bolshevism as a direct threat, both to the army and to the new federal government in Berlin. Hitler was chosen to remain in the army and attend a propaganda course in Munich on fighting Marxism and Bolshevism among the soldiers. He was then sent to lecture new recruits and returning German prisoners of war on the evils of the Soviet system. The lowly private first class (PFC) discovered he had a natural talent for public speaking and debating—which did not go unnoticed by his superiors.

Earlier in 1919, a right-wing locksmith named Anton Drexler and a journalist named Karl Harrer, and several others in Munich, formed an organization called the German Workers' Party, which was anti-Semitic, ultra-national, anti-Bolshevik, and opposed to the harsh terms of the Versailles Treaty, which required the new German (Weimar) Republic in Berlin to limit its army and navy and pay heavy war reparations. The new party also tapped into the anger and regret felt by many *Bavarians* over their loss of sovereignty, for which warmongering Prussia was blamed. Before the War, Bavaria not only had its own king and army, but also its own railroads, post and telegraph, and even its own foreign policy.

PFC Adolf Hitler, now thirty, still in the army, was assigned to spy on the new movement and attend its meetings. He identified with the ideology of the new little "party" (fifty-four members), and his debating skills impressed Anton Drexler, who invited him to join.

Also impressed by Hitler, was a morphine-addicted, anti-Semitic poet, playwright, journalist, publisher, and party member named Dietrich Eckart, who took Hitler under his wing and introduced him to a number of prominent right-wing patrons, both in Munich and Berlin. Like the other members of the Party, Eckart believed Jews were intent on world domination, a popular sentiment at the time.

In 1920, when the Nazi party was in its infancy, the world's most notorious and damaging anti-Semitic book, known in English as *The Protocols of the Elders of Zion*, first printed in Russia in the early twentieth century, was published in both German and English. A fictional product of the Tsar's secret police, it purported to be the minutes of the meetings of a Jewish plot to use private banking, capitalist businesses, mass media, the courts, labor unions, school systems, left-wing political influence, and even the Christian clergy to take over the world. In the US that same year, 1920, Henry Ford's *The Dearborn Independent* began running a series of anti-Semitic articles eventually published later in the year as

The International Jew, which often cited the *Protocols,* accused Jews of warmongering and profiteering, and was also instrumental in spreading anti-Semitism throughout the world. Hence, from the earliest days of the Nazi Party, Adolph Hitler's anti-Jewish rants had a very receptive audience. Jews were blamed for sponsoring both *capitalism,* which supposedly exploited the working man, and *Marxist-Bolshevism,* seen as a threat to the German economy and society in general.

Dietrich Eckart, who died in 1923 and was twenty-one years older than Hitler, had a great deal of influence on Hitler as a public speaker—urging him to appeal not only to working men but also to World War I veterans. With his theater background, Eckart helped Hitler learn *Bühnendeutsch,* or "stage German," an artificial pronunciation of written Standard German, developed by theater troupes, that gave Hitler more universal appeal than speaking in his native Central Bavarian dialect. Eckart became the first editor of the Party newspaper and was instrumental in marketing Hitler and his speeches, which were becoming the major source of revenue for the new Party. (At the end of the second volume of *Mein Kampf,* Hitler paid special tribute to Eckart).

By the time troop limitations in the Versailles Treaty forced Hitler's demobilization from the army in May of 1920, he had become the new party's star attraction. Earlier in the year, he was made its head of propaganda. Shortly thereafter, the right-wing organization changed its name to the National Socialist German Workers' Party (i.e. Nazi Party): "National" to make it sound less Bavarian, and "Socialist" in its willingness to nationalize some areas of the economy—Hitler was no real socialist. The movement now attracted thousands—many of whom were veterans—to hear speeches by the former PFC, and to make political donations. Hitler soon became the Party's leader.

While the Kaiser had been the head of state during World War I, the *de facto* military dictator of Germany during the War was General Erich Ludendorff, right-hand man of Field Marshal Paul Von Hindenberg. In the wake of Germany's war loss, Ludendorff claimed Germany had been "stabbed in the back"—that the defeat in the Great War was not the result of military failure (for which he bore responsibility) but because of lack of support on the home front by Jews, liberals, Communists, Socialists, striking unionists, and war profiteers—defeatists who had tired of war.

General Erich Ludendorff with Adolf Hitler (*Bundesarchiv Bild*).[5]

In the early 1920s, Ludendorff threw his support behind Hitler's fledgling Nazi Party, which took up the "stabbed-in-the-back" myth. The World War I military dictator mentored the man who was to become Germany's military dictator in World War II *and elevated his political status.*

In November of 1922, Germany missed paying reparations required under the Versailles Treaty. In response, in January of 1923, France and Belgium invaded and occupied Germany's industrial heartland, the Ruhr Valley, which resulted in a political crisis, and an economic depression in Germany with high unemployment and hyperinflation.

In November of 1923, in the midst of this economic and political unrest—and inspired by Mussolini's successful "March on Rome" the year before—Hitler and Ludendorff participated in the so-called "Beer Hall *Putsch*" in Munich, an attempted *coup d'état,* in which several were killed and wounded. Hitler was arrested and tried for treason and the Nazi Party banned. The headlines surrounding the *Putsch* and his trial and conviction gave Hitler notoriety throughout Germany. During his nine months at Landsburg Prison in Bavaria, he dictated the first part of his racist, almost unreadable, autobiographical manifesto called *Mein Kampf* that made him even more notorious.

While Hitler was still imprisoned, Germany sought to banish him—to deport him back to Austria. However, Austria, noting his service in a foreign army, refused to accept the political rabble-rouser—now a convicted felon. After his release on parole in December of 1924, and hoping to perhaps become a German citizen, Hitler, on April 7, 1925, renounced his

Austrian citizenship. To further his efforts at German citizenship, and in a successful effort to have his party reinstated, he promised Bavarian authorities, as a condition of parole, the Nazi Party would renounce violence and use only peaceful means to further its political goals. Germany, however, did not grant him citizenship; and in 1925—as he tried to rebuild his party—he was officially *stateless*, a man without a country.

In August of 1924, while Hitler was in prison, the Allies adopted the American-sponsored *Dawes Plan*, which restructured German war reparations and called for withdrawal of French and Belgian troops from the Ruhr, ending the political and economic crisis that had benefited the extremist Nazis. In February of 1925, shortly after his release from prison, Hitler made a speech Bavarian authorities considered an incitement to violence, and Hitler was banned from making any public speeches until 1928. Nevertheless, he succeeded in restructuring the Party, trying to spread it outside of Bavaria—but it remained on the fringe until political chaos came again to Germany.

After the 1929 US Stock Market Crash, the Depression spread to Germany. Businesses failed, unemployment soared, savings evaporated, banks collapsed—and Germany once again was in crisis. The traditional political parties were deadlocked, and the German people looked to radical parties, like Hitler's, for relief. In 1930, the Nazi Party became the second largest in the Reichstag.

In 1932, with unemployment approaching 40 percent, Hitler decided to run for President—but he was not even a German citizen, which was required under Germany's constitution. Hitler sought help from a Nazi official, in a local coalition government, who had become interior minister of the State of Braunschweig. Under a loophole in German immigration law, citizenship would be given automatically to someone appointed to a particular bureaucratic position in Braunschweig; and Hitler was given the post—at which he, of course, never worked. By this loophole, Adolf Hitler thus became a German citizen *at age forty-two*—only a month before the presidential election. (He was listed on the ballot as a Braunschweig *senior civil servant*).

During the 1932 campaign—which involved an initial election in mid-March and a runoff four weeks later—Hitler feared he was losing his voice. On the advice of a doctor, he engaged an opera tenor called

Paul Devrient to advise him as a voice coach and speech teacher to improve his "stage German." Despite vigorously campaigning, he lost to bigger-than-life (six feet, six inches), doddering, incumbent eighty-four-year-old Paul Von Hindenburg, who despised him. But after federal elections in July and November made the Nazis the largest party in the Reichstag, Von Hindenburg begrudgingly made Hitler *Chancellor* the following year. Unlike Mussolini, who owed his position to the King of Italy, Hitler, eleven years later, came to power, indebted to no one—that is, after Hindenburg died in office at age eighty-six in 1934, right after the constitution had been amended to make Chancellor Hitler both head of state and head of government on Hindenburg's death.

The two Axis countries, Germany and Italy, were very different. Germany was much larger and richer in natural resources. Italy had few natural resources, little coal or iron ore, had much less industry, a much less prosperous economy, and was mostly hills or mountains, except for some coastal areas and the Po Valley in the north.

Under Hitler, Germany went from becoming a republic to a *Reich*, or empire, with the *Führer* as an absolute ruler. Under Mussolini (*"Il Duce"*) the Kingdom of Italy, while *de facto* a dictatorship, legally remained a monarchy, modeled originally on the British system, with the head of state, the king, only five feet tall, who had reigned since 1900, having limited authority—Fascism's Grand Council even having taken from him the power to appoint *or dismiss* the head of government, the prime minister. In Germany, soldiers took an oath of allegiance to Hitler. In Italy, soldiers still swore allegiance to the king, and not to the powerful, charismatic prime minister, Mussolini.

Hitler had been a PFC and Mussolini a corporal in World War I, and both had been wounded at the front, Hitler *twice*—first by shrapnel in the leg, and later by mustard gas, and Mussolini in the leg and elsewhere in an accidental explosion—fatal to several others—of a weapon described as a *trench mortar or grenade launcher.* While both men were spellbinding orators, Mussolini, who was more educated, and a former journalist, was also an excellent writer, whereas Hitler had no skill in that area. In private conversation, translators later said, Mussolini came to the point, whereas Hitler tended to ramble.

. . .

On July 19, 1943, while the invasion of Sicily was still going on, Hitler, concerned about the deteriorating Italian situation, came to Northern Italy and met with Mussolini at a villa outside of a town called Feltre, north of Venice. Hitler spoke no Italian, but Mussolini, who had lived in Switzerland and Austria earlier in his life, spoke passable German (as well as French—and even some English). Italy seemed unable to defend itself, and the *Führer* was trying to decide what to do when the Allies invaded. Some of his advisors were arguing he should *not* open a third front in Italy and instead focus his resources on the Eastern Front and the anticipated invasion of France in the west.

Others, among them Field Marshal Albert Kesselring, argued giving up Italy would allow Italian airfields to be used to bomb Southern Germany, as well as the Balkan countries to the east, and Ploiesti in Romania, the most productive oilfield in Europe and main supplier of petroleum for the German Army and the *Luftwaffe* air force. Kesselring also advised that Italy with its eight-hundred-mile-long Apennine Mountains could be readily defended, and a German army could be easily supplied through German-annexed Austria and by Italy's heavy industry and rich agriculture in the north. The Po Valley, Italy's "breadbasket" was also seen as a major food supplier for Germany itself.

At Feltre, Mussolini agreed to let the Germans take over the defense

Mussolini and Hitler at Feltre. (*IstitutoLuce*).[c]

Victor Emmanuel III[c]

of Italy—essentially turning over command of the Italian armed forces. Hitler put Field Marshal Kesselring in charge. The same day Hitler and Mussolini met, the Allies bombed Rome, adding to Mussolini's unpopularity among the Italian people and emboldening his enemies in the government to get rid of him for having "Germanized" Italy.

On July 25, 1943, the Grand Council of Fascism, the heart of Mussolini's government, which had not met formally in four years, held a *no-confidence* vote. By 19 to 8, with one abstention, the Grand Council (secretly supported by the king) voted to restore the king's full constitutional powers—which included the power to discharge Mussolini—in effect, deposing the dictator. One of the votes against Mussolini was cast by his ambitious son-in-law, former foreign minister Galeazzo Ciano—later to pay for his vote with his life.

The next day, a desperate Mussolini met with King Victor Emmanuel III to argue that the alliance with Germany was Italy's best hope against the Allies. Instead, the king—fearful the monarchy itself was in jeopardy for having backed Mussolini—told Mussolini Italy was defenseless against the Allies with their air power, and that the prime minister, having lost popular support, was now "the most hated man in Italy." To his face, Mussolini was told he was *dismissed* as prime minister and replaced by seventy-one-year-old Marshal Pietro Badoglio.

As Mussolini left the palace, he was arrested and briefly exiled to the islands of Ponza and then La Maddalena. Eventually, he was taken to a mountain hotel hideaway at Gran Sasso, in the Central Apennines east of Rome, where he was guarded by two hundred *carabinieri* police.

In August 1943, with the Allies advancing, the Germans under Field Marshal Kesselring made a brilliant tactical retreat from Sicily—the eventual launching pad for the Allied invasion of the Italian mainland.

Italy was to be invaded by *two* Allied armies under the oversight of the 15th Army Group, then commanded by British Field Marshal Harold Alexander. The veteran Eighth Army under British General Bernard Montgomery would land in Reggio Calabria, across from Sicily and generally drive up the eastern, or Adriatic coast of the peninsula, while the newly-organized Fifth Army, under command of US Lieutenant General Mark Clark, would land at Salerno, within air cover from Sicily, thirty-five miles south of the *main objective*: the heavily-fortified port of Naples—and generally go up the western, or Tyrrhenian coast. Meanwhile, the new

Italian government, tired of war and now rid of Mussolini, was pretending to be loyal to the Germans—while secretly negotiating with the Allies.

All of this was somewhat anticipated by the Germans who were not totally surprised when the British landed on September 3, 1943 in the toe of Italy—though the Germans were surprised when King Victor Emmanuel III and the government announced an *armistice* with the Allies on September 8 and fled south from Rome to territory in Eighth Army control. With Italy switching sides, the US-led Fifth Army, hoping for only light resistance in its first enemy encounter, was initially stymied by the heavy German presence when it hit the beaches at Salerno, south of Naples, September 9, 1943.

Three days later, German paratroopers made a daring raid on Gran Sasso, rescuing Mussolini, who was flown to meet with Hitler in East Prussia (in today's Poland). The rescue was part of a German plan to capture Rome with German troops already in Italy and generally invade the country from the north at the same time the Allies were invading from the south. German *Waffen SS* (or *Schutzstaffel*) combat troops confronted Italian units and demanded surrender or cooperation.

Some Italian troops that remained loyal to the king and the new government of Badoglio were disarmed by the Germans—many Italian soldiers becoming prisoners, or slave laborers. Other Italian units continued fighting for a new Nazi-Fascist puppet government headed by Mussolini in Northern Italy called the *Italian Social Republic*. Still other troops, loyal to the king, mostly in the South, became part of the new Italian Co-belligerent Army, which fought beside the Allies, along with a Co-belligerent Navy and Air Force.

On October 13, 1943, under Allied insistence, Badoglio's new Italian government declared war on Germany.

Part of the craziness of the Italian Campaign was that uniformed Italian army units were fighting *on both sides*. In addition to uniformed soldiers, Italian civilians called *partisans*, affiliated with six banned political parties, like the Socialists and Communists—many of them disgruntled veterans—were armed by the Allies and performed intelligence and sabotage missions behind the lines, sometimes coordinated through American intelligence agent Max Corvo of the US Office of Strategic Services, or OSS. The partisans were also charged with trying to preserve infrastructure, like ports, bridges, factories, and power plants—for use when the Germans were ultimately defeated.

Just as the Allies had two armies fighting in Italy—the Fifth and Eighth, under the 15th Army Group—so did the Germans.

German Army Group C consisted of the Tenth and Fourteenth Armies, Army *Wehrmacht* troops, bolstered by the *Waffen SS*, ground forces—technically independent of the *Wehrmacht*, but which fought under the German Army command structure. Both sides, Axis and Allies, were at relatively equal strength for most of the War in Italy.

Given the needs of war production at home, Britain, operating its large wartime navy and merchant fleet, was desperately short of manpower and solicited fighting men and military resources from the Commonwealth countries and, along with the Americans, from countries that had been attacked or conquered by the Axis powers.

According to former 15th Army Group commander General Mark Clark, fighting in the Fifth and Eighth Armies next to the Americans and British, were Canadians, New Zealanders, South Africans, Indians, Swazis, Bechuanas, Seychellois, Mauritians, Rodriguez Islanders, Caribbean Islanders, Cypriots, Greeks, Lebanese, Jewish Palestinians (the Italians had bombed Haifa), Yugoslavs, Nepalese, Belgians, Poles, Free-French and their colonials: the Algerians, Tunisians, Moroccans, and Senegalese, the aforementioned Italians who switched sides (re-armed by the Allies), Australians who flew fighter missions, the Dutch navy—and *Brazilians*, among others.

Neutral Brazil was a major source of rubber, bauxite, and other raw materials and products needed by the Allies. After German and Italian U-boats sank over thirty Brazilian ships, including a troop ship and a transport full of pilgrims—with hundreds of lives lost—Brazil declared war on the Axis powers in 1942.

Reflecting Brazilian society, the racially-integrated *Brazilian Expeditionary Force*, with its fighter squadron and 1st Brazilian Infantry Division, under General João Baptista Mascarenhas de Morais, equipped by the Americans and organized like an American division, went to Italy in late 1944 to help alleviate Fifth Army's shortage of troops, reaching full strength just before Dad's unit was attached to it.

Skeptics in Brazil thought their soldiers and airmen would never make it to the Old World before the War ended. "When will our fighting men make it to the front?" was a common question; to which the response in Portuguese was *"a cobra vai fumar,"* which literally translates as "when the snake smokes." The equivalent English expression might be "when pigs fly." Upon landing in Italy, the Brazilians noticed American divisions had shoulder patches and requested one for the Ist Brazilian Division. In a popular show of whimsy, the patch finally approved was of *a snake smoking a pipe*.

The Brazilians had made several forays into the Apennine Mountains in late November and December, 1944, and had taken some key heights, including Mt. Belvedere, only to be later repulsed by fierce German counterattacks.

Generally, due to the need to build up stocks of ammunition, 105mm howitzers at this time were limited by Fifth Army to firing no more than fifteen rounds per day. Nevertheless, in support of the Brazilians, Dad's battery, over the course of a few days, had a number of fire missions, knocking out several enemy targets—an ammunition dump, a gun position, and a mortar position, among others.

That is how Dad's artillery battery—with Dad, I am sure, trying to get all 113 men a hot meal—got to spend Christmas of 1944 in support of the Brazilians, dug-in near the little northern Apennine mountain town of Capugnano, where the first winter snows had already fallen.

11

INTO COMBAT

Any illusions Dad may have had about sitting out the winter in the Campugnano/Porretta Terme area until the anticipated spring offensive were suddenly shattered on December 27, 1944. Dad's entire battalion was detached from the Brazilians and ordered west, out of the mountains, to defend the *coast* in support of the 92nd Division. While some elements of that division were along the coast, others were engaged in heavy fighting in a mountainous area of northwest Tuscany called Garfagnana, north of Lucca, where the Apennines met a coastal range called the Apuan Alps. Dad's battalion dug in outside of a seaside town called Forte dei Marmi, in a narrow coastal plain at the foot of the mountains, north of the beach city of Viareggio.

The 92nd Division was the westernmost unit facing the Gothic Line, the German line of defense, which went from the mountains all the way down to the sea. At the coast, the line was at the Cinquale Canal, today's Versilia River, only a few miles from where Dad's battalion was stationed. The other side of the Cinquale Canal was heavily mined and fiercely defended. While some units of the 92nd Division were along the coast, others had made advances earlier in the month by taking small towns in the mountains above.

The US 92nd Division, also called the *Buffalo Soldiers*, was African-American—including company-level and a few battalion-level officers, while most senior officers were white. One regiment of the division had been in Italy many months. Other regiments were newer troops assigned to help Fifth Army's manpower shortage. In the spring, the coastal troops were expected to advance past the Cinquale Canal toward the coastal town of Massa, which had maybe fifty thousand residents and was within artillery range of La Spezia, farther north, an Italian Navy port with heavy coastal guns used by the Axis. Beyond that was Italy's biggest seaport (and home of Christopher Columbus), Genoa, which served Milan and Turin. In the mountains to the east, other 92nd Division troops were positioned in several small towns in the Serchio River Valley, in what heretofore was considered a relatively quiet area of the front. The Serchio River ran north-south and separated the coastal mountains from the Apennines.

In northern France at this time, the Germans had launched an ill-fated offensive through the Ardennes Forest intended to surround the Allied armies and take the Port of Antwerp, Belgium, the main source of Allied supplies. Known in the West as the "Battle of the Bulge" the German plan was so desperate it depended on capturing Allied fuel dumps because the Germans did not have enough fuel for their advancing tanks.

The much-bombed Ploiesti oil fields and their refineries in Romania, the main source of petroleum fuel for the Nazis, had been overrun by the Soviets in August 1944, leaving only a smaller field in Hungary as a source of Axis oil. Although Germany built plants to make synthetic fuel from coal, the plants were frequently the target of Allied air raids. The quantity of synthetic fuel produced was insufficient for any large offensive operations, or so the Allies believed, and the quality was generally inferior for sophisticated uses, such as for aircraft.

Octane-boosted, petroleum-based aircraft fuel enabled Allied aircraft to have more powerful supercharged engines and longer range than those of the *Luftwaffe*. Hard to believe today, but the US, when it entered the war, was by far the world's largest oil supplier—generating more petroleum than the next ten oil-producing countries *combined*. The German

Army, in light of inadequate fuel reserves, relied heavily on horse-drawn wagons for supplies and largely relied on *horses* to reposition artillery.

With "The Bulge" going on in France, the Germans planned a similar desperate offensive in western Tuscany, having the objective of taking the Allies' main supply port in Northern Italy, Leghorn (*Livorno*), either by descending from the mountains and taking Lucca, or by attacking along the coast. However, it was soon apparent an attack from the Gothic Line south would have to be more moderate in scope due to lack of resources. At the same time, Italian troops loyal to Mussolini and his new socialist republic wanted to go on the offensive. A more moderate objective among the Axis planners was the hope a surprise attack in western Tuscany would move Allied reserves away from the center of the Gothic defensive line, where they feared an attack on the major transportation hub of Bologna in the Po Valley.

On December 26th, about 9,000 Axis troops—two-thirds of them Italian—and 100 artillery pieces attacked the American 92nd Division and routed them, capturing about ten mountain towns, which were looted for food, ammunition, fuel, and equipment; and 250 prisoners were taken. The Allies moved reserve troops, in this case two brigades of the 8th Indian Infantry, into position. Other troops, like Dad's artillery battalion, were moved to the coast to prevent any Axis advance toward Leghorn from along the coast.

On the afternoon of December 27, while Dad's battalion was on its way down from the mountains to the coast, the Axis objectives in the western mountain sector had mostly been met. Under pressure from the Indians, the enemy soldiers in the mountains began retreating, and eventually set up about a mile or so south of their original Gothic Line positions. However, pockets of resistance remained as the Indian troops attempted to regain lost territory in the mountains. Along the coast, artillery and mortars were set up by the Axis troops for firing on Allied troops, like Dad.

The mention of Indian troops brings to mind something Dad told me. Among the soldiers from various countries fighting on the Allied side were special mountain warriors from Nepal called *Gurkhas*, attached to Indian units. They supposedly had knives that could only be resheathed if blood had been drawn. So the other Allied soldiers would

Gurkha with kukri knife.[c]

ask to see the knife, called a *kukri*; and before a Gurkha could put it away, he would cut himself. That is what Dad told me—though others say the drawing-of-blood requirement was a myth.

Nevertheless, the Gurkhas were excellent soldiers often used for their shock value. Dad told me they were known to sneak into German camps at night, silently kill a sleeping soldier or two with their knives and leave as quietly as they came. Such exploits had an obvious demoralizing effect on the enemy.

French colonial troops from Morocco called *Goumiers* or *Goums* similarly had knives that supposedly could not be resheathed unless blood had been drawn and also were known to attack the Germans while they slept in their cots. By the time Dad got to Italy, the French and their colonial troops were mostly gone, part of Operation Dragoon, the invasion of the south of France; but Dad likely encountered Goums, and legends about them, in North Africa and on Corsica.

12

A CASUALTY

The soldiers with the most dangerous jobs in an artillery battery were the forward observer and his radio operator. The forward observer was a second or first lieutenant who worked with the infantry units. When he spotted a target, he radioed it in to the fire direction center, which did the mathematical calculations and gave commands by radio or field telephone to the guys manning the howitzers, the cannoneers. The enemy wanted nothing better than to take the forward observer and his radio operator out of action.

On the night of December 29, 1944, a first lieutenant forward observer from Dad's battalion, working with a 92nd Division Buffalo Soldier infantry regiment, came under heavy artillery and mortar fire and was seriously wounded, Dad's battalion's first casualty. He was not to be the last.

Italy saw an amazing event on January 1, 1945. With two P-38 Lightning fighters flying overhead for security, Fifth Army and Twelfth Air Force, complete with pads, helmets and uniforms, played American football against each other in a giant soccer stadium in Florence in front of some 25,000 fans. (Fifth Army won 20–0). Unfortunately, Dad had

A forward observer post for Dad's battery, A. (*Dad's photo, dated 1944*).[c]

Most of Dad's photos are undated. This one, however, had a date of January 7, 1944 and shows the direct hit the day before on one of the M7 Priests in Dad's battery. The equipment was quickly replaced, bringing Dad's battery back up to six M7s. (*Dad's photo*).[c]

to miss the so-called "Spaghetti Bowl" as his battery was in a dangerous forward area on the coast near Viareggio firing at the Germans.

One of the worst things an artillery battery can experience is what is called counter-battery fire. A constant objective of artillery is to knock out the enemy's artillery. That is why artillery batteries, like Dad's, would try to conceal themselves with camouflage, or otherwise.

L4 Piper Cub two-seat "Grasshopper" observation plane.(*Signal Corps*).[c]

Dad's battery got a dose of counter-battery fire in early January. On January 5, 1945, at about 5 p.m., enemy artillery rounds began falling around Battery A. Dad and the other soldiers in the battery dove into their foxholes, scared to death.

Dad's battery area came under another barrage the early afternoon of January 6, injuring four men. This time a direct hit was scored on one of the six M7 Priests in Dad's battery—fortunately, unmanned at the time.

The Germans soon were about to experience some counter-battery fire, too. While the men in Dad's battery were coming under this blistering attack, they had a friend in the sky. Headquarters battery of Dad's battalion operated two Piper Cub L4 Grasshopper observation aircraft, and one was airborne at the time Battery A was attacked on January 6, 1945.

The L4 Grasshopper was a two-seater—room for a pilot and a forward observer who could identify targets and relay coordinates by radio to the fire direction center. In this case, the observer was able to spot two active enemy firing positions, which were then taken out by Battery B—and the shelling on Dad's battery area stopped.

The Germans quickly learned not to fire on observation aircraft—for to do so only made one a target for the deadly American artillery.

13

CHANGE IN COMMAND

The enemy attack in the mountains above Lucca and the increased threat along the coast—where Dad was feeding the Battery A troops (when he wasn't in his foxhole)—presented the first challenge to a new Allied command structure put in place in December, 1944.

It was all precipitated by the untimely November death in Washington, DC of Britain's main general coordinating with the Americans, Field Marshal Sir John Greer Dill, who died of a blood disorder. Though not a member of the American Armed Forces, the much-respected Dill was buried with full military honors at Arlington National Cemetery. He was replaced by Field Marshal Sir Henry Maitland "Jumbo" Wilson, who had been Supreme Allied Commander in the Mediterranean. In turn, Wilson was replaced as Mediterranean commander by Field Marshal Harold Alexander, called "Earl Alexander of Tunis," who had been in charge of 15th Army Group, the top commander in Italy over both the US-led Fifth Army and the British-led Eighth Army. Fifth Army commander US Lieutenant General Mark W. Clark, viewed as a diplomat-soldier, like Eisenhower, was moved up to commander of 15th Army Group—in

15th Army Group commander Gen. Mark W. Clark, and Fifth Army commander Lt. General Lucian K.Truscott, Jr.[c]

Maj. General Willis D. Crittenberger, IV Corps commander.[c]

charge of all Allied troops in Italy. US Lieutenant General Lucian K. Truscott, Jr., an aggressive, hands-on commander, was put in charge of Fifth Army. He had commanded VI Corps when it was sent from Italy to France. VI Corps was replaced in Fifth Army by newly-created *IV Corps* under US Major General Willis D. Crittenberger.

Dad's battalion was assigned by Fifth Army to IV Corps, which was responsible for the western seventy, or so, miles of the ninety, or so, miles of Fifth Army's sector of the front. (The remaining heavily-defended twenty or so, miles of Fifth Army's sector, east of IV Corps, was assigned to the other corps in Fifth Army, II Corps, commanded by US Major General—later, Lieutenant General—Geoffrey Keyes, a West Point classmate of General Crittenberger's).

Meanwhile, the British-led Eighth Army, responsible for the eastern half of the front, had new leadership, too. Field Marshal Bernard Montgomery had been replaced by General Sir Oliver Leese when Montgomery was assigned to the Normandy invasion. At the end of December, 1944, Leese was sent to Asia; and General Sir Richard L. McCreery was given command of the remarkable Eighth Army, which had been successfully fighting the Nazis and Fascists since 1940, first in the Egyptian and Libyan deserts, then in Tunisia, Sicily, and now Italy.

So going into 1945, a whole new group of Allied generals was trying to end the Italian Campaign as soon as possible by getting through the

Northern Apennines, crossing the Po Valley, and cutting off the twenty-five German and five Fascist divisions they were facing—before the enemy could get to the Alps and set up another line of defense.

However, not a lot could be done without improvement in the manpower situation and in the winter weather—which hampered activities on the ground and in the air. The Allies were also short of supplies, including artillery ammunition, diverted to the other theaters.

A name I recall Dad mentioning several times, though I do not recall the context, was *Clare Boothe Luce*. One of the most influential women in America, a social-climber, journalist, magazine editor, successful playwright, and wife (in an open marriage) to *Time* and *Life* owner Henry R. Luce—she had been reelected as a Republican member of Congress from Connecticut and visited Fifth Army in Italy, after the change in command, as part of a delegation concerned about the "forgotten front." Feigning illness, she remained ten days after the rest of the delegation left for the States—and supposedly had a brief affair with the new commander of Fifth Army, General Truscott.

Writing and speaking about Fifth Army's needs, she became its champion in Washington, and the champion of its ordinary soldiers—advocating more R&R (rest and recuperation) and "combat" or "hazardous duty" pay. Subsequently, in the Eisenhower administration, she was named Ambassador to Italy, the first woman named a foreign ambassador in our country's history.

Rep. Clare Boothe Luce (R-Conn.) with 15th Army Group Commander General Mark W. Clark in Dec., 1944. (*Signal Corps*).[c]

The stalemate in the Apennines—while the Allies were building up their forces and awaiting new stocks of ammunition and supplies—gave the Germans additional time to fortify their Gothic Line, with its many anti-personnel and anti-tank mines, and over two thousand reinforced mountain bunkers, pill boxes, tank traps, machine gun and artillery emplacements, many blasted out of the rock or constructed of reinforced concrete by the fifteen thousand slave laborers put on the project that lasted over a year—engineered by the notorious German civil and military engineering group *Organization Todt.*

While the Allies' main thrust over the mountains and toward the Po was expected now to come in spring, Dad's battalion would see a lot of fighting before then.

Fifth Army's new IV Corps had a lot of untested troops.

Two limited, division-size offensives were authorized for Fifth Army during the winter, both in the IV Corps sector—Dad's sector—and Dad would see action in both of them.

14

THE BUFFALO SOLDIERS

The performance of the all-black 92nd Infantry Division at the Battle of Garfagnana in December of 1944 troubled Fifth Army's leaders. Troops that were expected to stand and fight were described as "melting away," particularly during nighttime. Had the Axis soldiers attacked where they did, in the 92nd's sector, because they sensed a weakness among the black troops?

The United States military during World War II was racially segregated. Integration did not start until three years after the War ended. The Pentagon building, constructed during the War in accordance with Arlington County, Virginia, building codes, was actually completed with separate "white" and "colored" lavatories—though President Roosevelt forbade any such labels in the federal facility.

African-American leaders in Washington had agitated for a greater role for blacks in the war effort, threatening marches, for example, if black businesses did not get more of the lucrative wartime defense contracts. High numbers of blacks were rejected for military service, and President Roosevelt made an order that African-Americans should be accepted into the military in numbers comparable to their percentage in the overall population (about 10 percent). Yet in the segregated schools of the Jim Crow South and in inner cities in the North, African-Americans received inferior educations. Though frequently attacked as racially biased, the *General Classification Test*, an intelligence/aptitude test given to new recruits, categorized recruits into one of five classes following a bell curve, with Class I being highest and Class V being the lowest. Among white recruits, under 10 percent were in Class V, while nearly 50 percent of African-Americans fell into the lowest category. According to one source, 75 percent of the enlisted men in the 92nd Division were either in the bottom two classes or unable to take the test because they were illiterate—and very few were in Class I.

The 92nd Division trained at Ft. McClellan in segregated Alabama. African-Americans from the North, who had only known *de facto* segregation, suddenly learned they could not drink at the same water fountains as whites, or eat at the same restaurants.

Much effort went into training and equipping the three infantry regiments and four artillery battalions of the 92nd. They proudly wore the "Buffalo Soldier" patch, harking back to an African-American cavalry unit at Ft. Leavenworth, Kansas during the so-called Indian wars.

The question of whether African-American soldiers could fight seemed to have been answered by the arrival in Italy of the 370th Regimental Combat Team, later called the 370th Infantry Regiment, in early August 1944. The 370th, which trained for two years in Arizona, had rejected many soldiers with substandard scores and represented what was thought to be the best in African-American soldiers, with black officers at the company level and even a few at the battalion level. To the relief of the Army and their supporters back home, the 370th was making a significant contribution to the war effort. As part of Fifth Army, attached to the 1st Armored Division, the 370th did what was expected of it, with signs of individual heroism as the Germans retreated from Rome across the Arno and finally to the Gothic Line. This mostly involved mopping up and light fighting, but the 370th lost a number of its best officers and senior noncommissioned officers.

In October 1944 trouble began. While awaiting the arrival of the other two regiments of the 92nd Division, the 370th was placed under

92nd Division commander Major General Edward M. Almond inspecting troops. (*Signal Corps*).

command of newly-arrived division commander, Major General Edward Almond in Viareggio.

Hastily, he planned an attack on Massa, the western coastal town north of where Dad would later be, by initially taking the heights overlooking Massa, the first of which was Mt. Cauala. One after the other, units of the 370th were sent up the mountain and came back down, sometimes facing only light opposition, and at other times facing machine gun, mortar, and artillery fire. The excuses were many. Communications were bad. Units did not know what to do. When officers were killed or wounded, the whole unit quit. The black enlisted men did not trust their "Uncle Tom" black officers, and the black officers did not trust each other or the white officers over them. The enlisted men would not go on a "suicide mission," which was just about any mission that involved risk of harm. Although some units performed well, morale was low, and it was difficult to motivate the men when the units right and left were slipping away.

Meanwhile, the two other regiments of the 92nd Division, the 365th and 371st, arrived during October 1944, but were not yet combat-ready. In addition, another African-American infantry regiment, the 366th had been broken up and was doing guard duty at Air Force bases all over Italy. It was reassembled and assigned to the 92nd as a fourth regiment. A major concern was replacing black officers and senior NCO's who became casualties. The Buffalo Soldiers lacked a "deep bench."

Although General Almond believed the problems with the 92nd could be ironed out, Fifth Army commander General Truscott, especially in light of the poor performance of the division during the enemy's Christmas offensive in Garfagnana, wanted to test what the new and untried 92nd could do as a division. So, plans were made with limited objectives in the coastal area, where Dad's battery had been since right after Christmas. The artillery ammunition shortage was expected to be somewhat corrected by the end of January, so an offensive in early February 1945 was planned for the 92nd Division.

While the front may have been considered a stalemate, it was by no means idle. During the month of January 1945, after the loss and replacement of the M7 early in the month, Dad's battery—still on the coast southeast of Forte dei Marmi, near the Cinquale Canal—was busy with fire missions. On January 9th, they destroyed a bridge. On the 12th, they knocked out an enemy pill box. On January 13th, Dad's battery area was hit by five enemy incoming rounds. On the 18th of January, they were hit with twenty incoming rounds in and around the battery area. On the 19th, they fired in support of an infantry patrol. On the 22nd, they fired sixty rounds, hitting an enemy ammunition dump. The rest of the month, they fired a number of antipersonnel missions.

On February 2, in anticipation of the 92nd Division's advance up the coast, a company of tanks from the 760th Tank Battalion—I assume six in number—was temporarily attached to Dad's battalion and given the designation Battery X. Earlier in the campaign, Fifth Army had also used the guns of its tanks, usually 75mm, as a form of artillery to supplement the howitzers. Dad's battalion had been attached to the 92nd Division and was under the division's artillery command. On February 5, 1945, Battery A, Dad's battery, was inspected by the division, so all was ready to go for the offensive.

A diversionary attack in the mountains by elements of the 92nd Division began on February 5, 1945 with the main coastal attack, combining armor and infantry, coming on February 8. The battle plan called for tanks and infantry—Task Force 1—to go around the Cinquale Canal at the beach while infantry units of the 92nd Division—370th and 371st Regiments—broke through in the heights, which began four or five miles inland. In support, Dad's battery, already stationed just south of the Cinquale Canal, fired

Coastal Cinquale area of northern Tuscany in 1945. (Eastern part of canal to the right). (*Signal Corps*).

Cinquale area today. Massa was the objective. Dad's battery was near Forte dei Marmi. The Cinquale Canal marked the coastal area of the Gothic Line, Allies were to the south (right in photo) and Axis troops to the north (left of the canal in photo). (Google Earth view. ©*2017 Google*).

about 800 rounds against enemy personnel, mortars, and gun positions the first day, 500 rounds the second day, and 760 rounds the third day of the attack. By then it was clear the offensive was a disaster, both from a military and a political standpoint.

During the first day of the main battle, February 8, 1945, the sky was clear, and air support was excellent. The two regiments in the mountains made modest gains, about a thousand yards for the 370th—though the 371st, on its right flank, was stymied by a steep hill. The tanks and infantry on the coast also advanced north past the Cinquale Canal while losing several tanks due to mines. The infantry battalion there came

92nd Division troops advancing on Cinquale Canal. February 8, 1945. (*Signal Corps*).

under heavy fire; and during the night, according to Fifth Army Commander Truscott, as many as two hundred troops "melted away" and were collected as stragglers south of the canal.

The next day, February 9, it rained, and there was no air support. The battalion north of the canal was reinforced by two companies. The troops came under heavy artillery, mortar, and machine gun fire from the defending 148th German Infantry; and during the night straggling occurred as before, according to General Truscott. A number of troops supposedly abandoned their equipment and clothing and sneaked back south of the canal. General Truscott wanted to halt the operation after that, but granted the request of General Almond, the commander of the 92nd Division, to persist one more day. The Germans had counter-attacked in the hills, and continuing losses in the coastal sector prompted General Truscott finally to end the operation. All Allied units retreated by the morning of February 11, 1945, about seventy hours after the offensive began, to where they were before the attack.

Sixteen medium and four light tanks had been lost—as well as two M7 Priests by the 27th Armored Field Artillery Battalion. The infantry battalion that had made it past the Cinquale Canal lost two officers and

Google Earth view of a coastal gun emplacement at Punta Bianca today (minus guns). (©2017 *Google*).

31 enlisted men killed, and ten officers and 171 enlisted men wounded, with 48 enlisted men missing. The farthest advance was little more than a thousand yards.

A major factor in the defeat was the unforeseen enemy use of long-range coastal artillery guns in hardened bunkers guarding the Gulf of La Spezia at a place called *Punta Bianca* (White Point) some ten or twelve miles away. The coastal guns were guarded by at least six Flak anti-aircraft guns, and Allied aircraft were unable to take them out. Since they were beyond the range of Allied field artillery, the coastal guns were able to support the Axis troops, both on the coast and in the hills above, with impunity.

A few days after the offensive, Five-Star General George C. Marshall, Chief of Staff and highest US Army officer, stopped off in Italy on his way back to Washington from the Yalta Conference in the Russian Crimea, where Roosevelt, Stalin, and Churchill met for the last time.

General Marshall was responsible for running the entire Army for the US. He believed the American soldier, whose country called on him to give up his job, family, and possibly his life, should have the best of everything. The general had a soft spot for Fifth Army. Less than nine months before, his beloved stepson, Allen Tupper Brown, a 1st Armored

Task Force 1 going around the Cinquale Canal on the beach. February 8, 1945. (*Signal Corps*).[c]

General of the Army George C. Marshall, Army Chief of Staff, highest officer in the U.S. Army, being driven on an inspection tour of the 92nd Division in Italy by the division commander, Major General Edward M. Almond. General Almond's driver is sitting behind. February 14, 1945. (*Signal Corps*).[c]

Division lieutenant, while peering out of his tank with field glasses, had been killed by a German sniper.

Basically, Generals Truscott and Clark told General Marshall the 92nd Division had valiant soldiers and many excellent elements, such as armor and artillery; but due to a number of issues, including difficulty replacing its company-grade officers and senior NCO's, was not capable of effective offensive operations as an infantry division.

After Marshall inspected the Fifth Army front, including the 92nd Division, it was decided the best African-American troops should be consolidated into a single regiment, the 370th, and the other two regiments in the division would consist of a white unit called the 473rd, made up of former anti-aircraft personnel, like Dad, and a regiment including the famous 442nd Regimental Combat Team (Nisei) and the 100th Infantry Battalion, both comprised of Japanese-Americans, which would be brought back to Italy from France.

The all-volunteer Japanese-American regiment, was to become the most highly-decorated combat unit of World War II.

So, the proud all African-American 92nd Division ended up being totally diverse, jokingly called a "Rainbow Division," with a black regiment, a white regiment, and an Asian regiment. (The real Rainbow Division, the 42nd Infantry, fought on the Western Front).

By 1945, General Marshall had transformed an army of 175,000 in 1939, rated nineteenth in the world, into one of over eight million with the best equipment. By the end of the War, with twelve million men and women in uniform in all the services, the US had become a superpower.

A leader not only in war, but also peacetime, the much-respected General Marshall became Democratic President Truman's Secretary of State and convinced a Republican Congress of the need for Ameri-

Fleet Adm. Wm. D. Leahy
(U.S. Navy)

Weekly luncheon of the Joint Chiefs in 1943 (before 5-star rank). Left to Right: General Henry H. "Hap" Arnold (Army Air Forces); Adm. William D. Leahy (Chair); Adm. Ernest J. King (Navy); and General George C. Marshall (Army). (*U.S. Navy photo*).

can aid to rebuild the post-war economies of our devastated allies and enemies as a bulwark against Communist aggression. The Marshall Plan was instrumental in making our modern world order, and General Marshall was awarded the Nobel Peace Prize in 1953.

During World War II, General Marshall was outranked only by an unsung Navy man and former ambassador to Vichy France, Fleet Admiral William D. Leahy, President Roosevelt's chief military advisor, or "Chief of Staff to the Commander-in-Chief," the de facto Chairman of the Joint Chiefs.

Born in Iowa, Admiral Leahy grew up in Ashland, Wisconsin, where he graduated from high school. An Annapolis graduate and son of a University of Wisconsin alumnus, Admiral Leahy was awarded an honorary Doctor of Laws degree by the University of Wisconsin when he visited Madison for commencement exercises in May of 1943. A few days later, he was awarded another honorary Doctor of Laws degree by

Northland College in Ashland, Wisconsin, which he always considered his hometown.

Whatever the makeup of the 92nd Division, General Almond learned much from the February incursion, and his re-formed division ultimately was successful in taking the coastal objectives—Massa, Carrara, La Spezia, and Genoa—as part of the spring offensive a few months later.

African-Americans had fought well in all the nation's previous wars, usually in segregated units with white officers, and in units smaller than division-size. The Army in World War II wanted to make the best use of its African-American members, but could not determine why the infantrymen, as a whole—with notable exceptions—did not fight successfully. One camp blamed the white leadership of the units, involving a lot of allegedly racist southerners. Another blamed racism in the country as a whole and the lack of quality education for African-Americans—the Army being but a reflection of its citizens.

In any event, the 92nd Division's failure as an all-African-American infantry division in World War II was a factor in the subsequent determination by the War Department/Defense Department that racial segregation in the military was harmful to the national defense and should be ended—despite the anticipated backlash from Southern soldiers and politicians (and the minority of African-Americans who felt more comfortable in African-American units). Experimental integrated platoons on the Western Front had proven successful; and, of course, Fifth Army's fully-integrated Brazilian division showed the races could fight successfully side by side

After the War, in July of 1948, President Harry S. Truman issued Executive Order 9981 to prohibit racial discrimination in the armed services. An advisor to the President was the only black general officer in World War II, Brigadier General Benjamin O. Davis, Sr.

Integration was slow to happen, however. The order carried no timeline, and most of the racial integration of the military did not occur until after the Korean War, under President (and former general) Dwight D. Eisenhower in 1954—the same year the U. S. Supreme Court determined unanimously, in Brown v. Board of Education, the Equal Protection Clause of the Constitution forbade racial segregation in public education throughout the country.

The controversial test scores of African-American military recruits during World War II were used in court to demonstrate segregated public schools in the South had denied African-Americans equal educational opportunity.

The same equal protection rationale also applied to underfunded public schools in the North, in African-American neighborhoods segregated by racial restrictions in private real estate conveyances. (Racially restrictive clauses in deeds and land contracts were ruled unenforceable by the Supreme Court in an earlier civil rights case, Shelley v. Kraemer, in 1948).

Prior to World War I, about 90 percent of African-Americans lived in the South, and 90 percent of those in rural areas. Historians date the economic boom-time of World War I as the beginning of the Great African-American Migration from the rural South to urban areas in the North. World War II produced a second wave, or some say a second migration, as millions of African-Americans relocated during and after the War, not only to the North, but also to urban areas in the West to work in steel mills, automobile factories, meat packing plants, and for the busy railroads—among many other employments. This huge demographic shift, and the acknowledged contributions of African-Americans to the War effort, led, over several decades, step-by-step, to hard-won progress in civil rights after the War.

15

BACK TO THE MOUNTAINS

On February 14, 1945, a few days after the 92nd Division retreated, Dad's battalion was detached from the 92nd and ordered back into the Apennines to support Fifth Army's other major winter offensive. While the rest of the battalion was sent to a place called San Marcello Pistoiese in support of what was called Task Force 45, Dad's battery, using its own integral fire-direction center, was attached to the 175th Field Artillery Battalion in support of the third battalion of the 86th Mountain Infantry around Porretta Terme, near where Dad had been before going to the coast. Task Force 45 was a stop-gap infantry-armor unit made up mostly of former anti-aircraft artillerymen (like Dad) and was in the process of being removed from the front and reformed as the 473rd Infantry Regiment (for the reorganized 92nd division). Its place on the IV Corps front was taken by the newly-arriving *10th Mountain Division*, which was to play a big role in Dad's movements over the next few months.

Dad's move from the coast back to the mountains meant he would be drawing rations from a different depot. Supplied by the quartermaster

Open-air Fifth Army food ration dump, this one in a grove of pine trees. (*Signal Corps*).[c]

A Fifth Army mobile bakery. (*Signal Corps*).[c]

corps, virtually all of Fifth Army's food came from the United States, usually first to Naples and then to the Tuscan port of Leghorn. From there, it went by rail or truck to Army depots at the major Tuscan cities: Florence, Pisa, Lucca, and Pistoia. While Dad was on the coast, the food probably came through Lucca (once home of composer Giacomo Puccini). In the mountains, food for Dad's mess likely came through Pistoia. From these main depots, the food was transported by truck to what were called "dumps" in the forward areas. A ration truck from the supply and transportation platoon of the battalion's service battery would have to go to the closest "dump" to pick up rations for the hundred or so men in Dad's battery mess. If possible, the dump was in a building, but sometimes even in the open air. The dump system was used not only for food but also for fuel, ammunition, medical supplies, and engineering equipment.

A priority food item for Fifth Army was bread. The quartermaster corps set up a series of mobile bakeries in the Fifth Army area and made fresh bread and rolls with flour brought over from the States.

Food preparation was "by the book." In August of 1944, the War Department distributed a technical manual called Army Recipes, TM 10-412, which contained hundreds of recipes with quantities to serve one hundred men, the approximate size of a battery, like Dad's. Food from the States generally meant what was called B-rations, as opposed to A-rations which was fresh food. Basically, B-rations meant the food, made in the kitchen or on kitchen ranges in the field, was not fresh, since it had to come by ship across the ocean. TM 10-406 gave recipes for cooking dehydrated food.

Eggs, potatoes, onions, and many vegetables were dried or powdered. With no refrigeration, milk was powdered or condensed. Meat was dried, cured, brined, or canned—Dad was so sick of Spam he never ate it after he got out of service. Fruits were canned, and vegetables were canned, or dried, like beans. From B rations came the despised breakfast offering S.O.S. ("shit on a shingle"—creamed (dried) "chipped beef" on toast). A convenient way of distributing B-rations was the 10-in-one, a

Typical M and B unit C-rations.

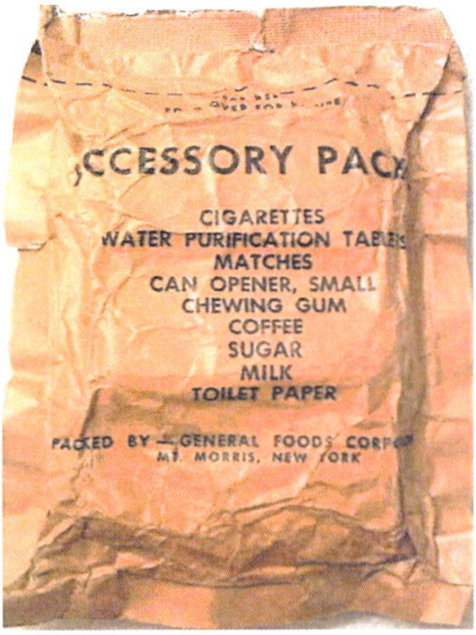

An accessory pack.

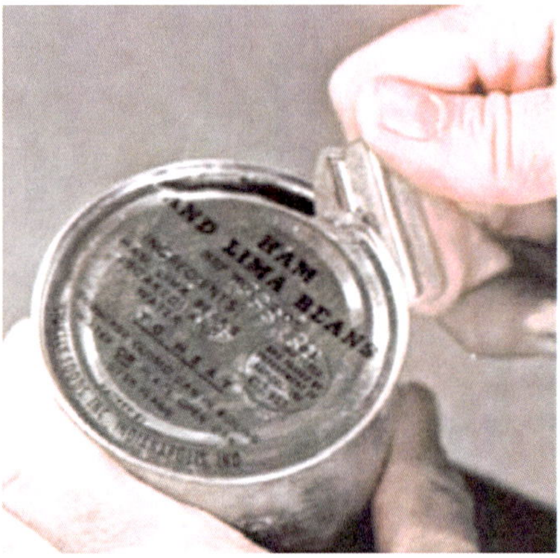

The P-38 can opener in action.[c]

single box that contained enough B-ration food to feed ten men three meals in one day.

If Dad and his kitchen truck could not provide hot food, the men in his battery were given C-or K-rations. A day's C-ration for one man consisted of six twelve-ounce cans of food—three cans of M-unit (for meat) and three cans of B-unit (for bread and other things). In addition, the soldier received one "accessory pack" per day.

Basically, the soldier was supposed to eat one M-can and one B-can at each meal. By the time Dad was in Italy, the C-rations had been improved with a lot more variety than earlier in the war. The M-cans had such offerings as: meat and beans; meat and vegetable stew; meat and spaghetti; ham, egg, and potato; meat and noodles; pork and rice; frankfurters and beans; pork and beans; ham and lima beans; and chicken and vegetables.

The B-units had a variety of different things in them, including biscuits, cereal, peanuts, raisins, instant coffee, lemonade and orange juice powder, cocoa beverage powder, jam, candies, caramels, sugar, and chewing gum. Some were of pound cake and such things. The accessory packs sometimes came in a package and sometimes in a can. They included such conveniences as a 9-pack of cigarettes, matches, gum, coffee, sugar, powdered milk for coffee—and toilet paper. The cans could be opened with a key, like a sardine can, or with a small foldable can opener found in the accessory pack called a P-38.

Together, the six cans were to provide about 3600 calories per day.

C oleman stove.

Fuel tablet heater.

C-rations could be eaten cold, but most soldiers heated the M-unit can on single-burner stoves invented by the Coleman Company that could use regular leaded gasoline. When not in use, the handy 9-inch high "pocket stoves" fit inside a compact aluminum capsule whose two parts could be used as vessels for heating food or water. Over a million were

K-rations were boxed as *Breakfast, Dinner, and Supper*. Note cigarettes and matches. (Dad first smoked Camels, then Old Gold brand until the 1960s when he switched to a filtered brand, L&M.) (*Signal Corps*).[c]

D-rations provided 600 calories, so three a day would provide enough calories for survival.

manufactured during the War. Alternatively, C-rations could be heated on little fuel-tablet "Sterno" type heaters.

As an alternative to C-rations, in situations where weight was an issue, like for paratroopers, and for harsh environments, the military developed K-rations. These came in three durable boxes, one for each meal and provided over 3700 calories. A day's ration of food weighed about two pounds. Each meal contained a canned meat or cheese product and biscuits. Included also was toilet paper, instant coffee, cigarettes and matches, and candy bars and desserts.

Finally, there were survival rations called D-rations. These were 4 oz. bars of a kind of chocolate intentionally made not to taste too good.

16

THE 10TH MOUNTAIN DIVISION

The other winter offensive for Fifth Army was designed to test the combat effectiveness of a newly-arrived IV Corps division, originally called the 10th Light Infantry Division and then renamed the *10th Mountain Division*. It had been turned down by the other theaters, but accepted by General Clark after Fifth Army lost the highly-effective mountain troops of French General Alphonse Juin to the Operation Dragoon invasion in Southern France. Fifth Army was in need of mountain fighters to get through the Northern Apennines. The 10th Mountain Division had trained extensively in the United States, especially at Camp Hale, Colorado, but had almost no combat experience. It was made up of a lot of mountaineers and ski enthusiasts, many recruited through the National Ski Patrol, which had lobbied the War Department for mountain troops. Drawn from the Cascades, Sierra, and Rockies in the West, and the Appalachians and Adirondacks in the East, the division had a high percentage of college boys and the highest test scores in the Army.

10th Mountain Division 75mm pack howitzer. (*Signal Corps*).[c]

However, as a division specially trained for mountain fighting, it was woefully light on artillery. Its three integral artillery battalions were armed with three 4-gun batteries of 75mm howitzers of the "pack" type. Pack howitzers were small, made to be broken down for transport and carried in pieces by six, or so, mules to get to otherwise inaccessible areas. However, such artillery lacked the range, firepower, and mobility of 105mm self-propelled M7 Priests like in Dad's unit. The 10th's artillery battalions were beefed up by Fifth Army. The three 4-gun batteries of 75mm pack howitzers were changed to two 6-gun batteries; and a battery of four towed 105mm howitzers was added.

Nevertheless, Dad's battalion was probably created with the intention of supporting the lighter artillery of the 10th Mountain Division, expected to be combat-ready in early 1945. (Shells for the 75mm pack howitzers were manufactured at the Milwaukee Works of International Harvester Company, where Dad's future father-in-law worked).

The commanding officer of the 10th Mountain Division was Major General George P. Hays, a slight, gawky, Oklahoman and former pack

Maj. General George P. Hays.[c]

artillery officer who was given command of the division in November 1944 when the prior commander became ill.

Despite his unimposing appearance, General Hays was a bold and brilliant strategist of unparalleled courage, and his men loved him.

As a young lieutenant in France in World War I, he had coordinated critical artillery fire by riding on horseback between command posts and artillery batteries after the Germans had destroyed the field phone lines. Over two days, he had seven horses shot out from under him and was severely wounded.

For this, he was given the nation's highest award for valor, the Congressional Medal of Honor.

General Hays was the architect of a plan for the 10th Mountain Division to take an objective known as Mt. Belvedere, the second limited offensive undertaken by Fifth Army in the winter of 1945, one of the most daring US attacks of the war—and Dad was a part of it.

17

HIGHWAY 64

Dad told me several times, "World War I was a war of railroads; World War II was a war of *highways*."

World War I involved mostly trench warfare with stagnant fronts that hardly moved during most of the war. Only near the end of World War I did the Germans and the Allies experiment with mobile warfare—the British using tanks to advance from the trenches across the barbed wire, and through the mines, artillery, and machine gun fire of no-man's-land.

Between the wars, the Germans, wanting to avoid the stalemate of trenches, developed a highly mobile type of warfare the Allies called Blitzkrieg—a term Hitler scoffed at—coordinating armor, air power, towed artillery, and mechanized infantry—all connected by radio communication. Highways and roads became critical not only for advancing armor, but also for fuel trucks, ammunition trucks, and other vehicles necessary for resupply, to keep the attack going.

Unfortunately, for the attacking Allies in Italy, the highways in the mountains could be easily defended, especially against armor. Fifth

Army had learned a painful lesson early in 1944 when it tried to go on the only highway left open from Naples north to Rome, Highway 6 in the Liri Valley. The German defenders at Monte Cassino, overlooking the highway, fought off three major Allied attacks, and stopped General Clark's Fifth Army advance toward Rome for four months.

The Germans were surprised by a flanking maneuver when VI Corps landed at the beaches of Anzio, southwest of Rome, behind the German line, designed to break the stalemate at Cassino. But when the Allies wasted time securing the beachhead instead of moving inland, the Germans were able to counterattack, and VI Corps, under constant artillery fire and subject to Luftwaffe air raids for months, could not break out from the coast. At Anzio, General Lucian Truscott, destined to be Fifth Army commander, went from being a division commander to assistant corps commander to commanding general of VI Corps.

When Fifth Army finally pushed past Monte Cassino, with help from French Moroccan mountain fighters and brave Polish troops, VI Corps simultaneously broke out from Anzio. Field Marshal Alexander, who was at the time in charge of both Allied armies in Italy, had given General Clark orders that VI Corps, breaking out from Anzio, was to sweep up the retreating German forces. Instead, General Clark, apparently fearful the English-led Eighth Army might steal from him the accolade of taking Rome, ordered General Truscott's VI Corps to head to the Italian capital, which allowed the retreating German troops—most of a whole German army—to escape and regroup. General Truscott could not believe the change of orders. This was a violation of one of the most basic tenets of warfare. You go after the enemy army, not after nonstrategic real estate. General Clark got what he wanted, the glory of entering Rome on June 4, 1944. General Truscott and most experts believed General Clark's actions prolonged the war in Italy for many months more than necessary.

After taking Rome, which had been abandoned by the Germans, General Clark pursued the enemy north in an effort to end the war in Italy by Christmas, but could not get into the Po Valley before weather and manpower and ammunition shortages resulted in Field Marshal Alexander calling a halt to offensive operations of both Fifth and Eighth Armies in the fall of 1944. By that time, General Clark's Fifth Army, still in the mountains, had gotten to within about ten or twelve miles of Bologna in the Po Valley on Highway 65, which ran from Florence to Bologna through the Savena River Valley in the Northern Apennines.

Meanwhile, General Truscott's VI Corps was sent to Southern France as part of Operation Dragoon. His success there led him to be named commander of Fifth Army effective December 16, 1944 when General Clark was promoted to head of the 15th Army Group in Italy and Field Marshal Alexander was made top general in the Mediterranean. At that time, the plan was to continue down Highway 65 in the spring offensive. Highway 65 was in the II Corps sector of Fifth Army.

There was another highway leading to Bologna through the Apennines, however, west of Highway 65. It was Highway 64, also called the Porrettana Highway, which originated in Pistoia and followed the Reno River Valley. The highway, on the very western edge of II Corps' sector, had been a pathway through the Apennines joining Tuscany with the Po Valley since pre-Roman times. The Reno Valley was also the location of the single-track Porrettana Railway. With numerous bridges, tunnels and viaducts, it was considered a tremendous feat of engineering when it was completed in 1864 to connect Bologna in the Po Valley to the spa town in the mountains. The Germans, of course, had sabotaged the railroad. They had also alienated the area citizenry when the SS, in fall of 1944, systematically killed at least 770 citizens—including women and children, and five Catholic priests—in the small Reno Valley town of Marzabotto, for collaborating with the partisans.

What was called a "highway" in the Apennines was not what most of us would think. Usually, it was nothing more than a two-lane road with no shoulders. The many towns and hamlets in the Apennines often had no sidewalks, and buildings came right up to the narrow "highway" as it passed through town.

When he took over Fifth Army, General Truscott determined the rough terrain in the IV Corps sector *west of Highway 64* was more lightly defended than Highways 64 and 65 in the II Corps' sector and might offer an easier route to the Po, with fewer casualties than the two highways—the defenses of which had been reinforced by the enemy over the winter. Truscott believed taking the mountainous territory west of Highway 64 in the IV Corps' sector would outflank the Germans and cause them to abandon their defense of both highways in the II Corps' sector, which could then accommodate the reserve 1st Armored Division and/or the 6th South African Armoured Division. (Fifth Army held its armor in reserve because the tanks were not suited to mountain fighting but would be necessary in the flat lands of the Po Valley once the planned spring breakout occurred). Truscott also figured obtaining the high ground west of Highway 64 would open up the possibility of using the less-defended Samoggia Creek or Panaro River Valleys on the north side of the mountains as a means of getting armor into the Po.

General Clark, however, refused to consider Truscott's attack idea. The plans had been made. When spring came, the main attack for Fifth Army would be in the heavily-defended II Corps sector up Highways 65 and 64—not in the mountains and hills of the IV Corps sector, Dad's sector, west of Highway 64.

However, at General Truscott's request, General Clark sanctioned a limited-objective thrust by Fifth Army west of Highway 64 in February 1945 to test the effectiveness of the new 10th Mountain Division and to try to extend Allied control of Highway 64 by about ten miles. The main objective was known as Mt. Belvedere, which had been attacked unsuccessfully three times before in November and early December, 1944, while General Clark was head of Fifth Army—once by the 92nd Division, once by the former anti-aircraft troops turned infantry, and once by the Brazilians. Each time, the Germans counter-attacked and kept control of the mountain, which had a commanding view of Highway 64 from which to call in artillery. From the heights, the Germans also were in position to view any preparations made by the Allies to attack the mountain.

Dad did not know it at the time, but the success or failure of the 10th Mountain Division's February 1945 attack on Mt. Belvedere was to determine where he would be until the war in Italy was over.

18

MT. BELVEDERE

Despite all of their training, the 10th Mountain Division troops never had to fight on skis. However, in January and February of 1945 some of the patrolling was done on skis. General Hays, division commander, needed all the intelligence he could get while planning to take Mt. Belvedere, which was really only one of the objectives of the offensive. He determined the reason the previous three attempts to take the mountain were unsuccessful was because of the German presence, and associated artillery observation posts, on a major outcropping of several peaks collectively called *Riva Ridge*, southwest of Mt. Belvedere. He wanted his men to take Riva Ridge with its 1,500 ft. sheer rock face, before attacking Mt. Belvedere and the heights east of Mt. Belvedere.

Generals planning and defending against attacks always count on reserves. Attacking generals use reserve troops to exploit any weakness in the enemy line, while defending generals keep troops in reserve to plug any gaps in the line that develop. General Hays was concerned that if the Germans saw his whole division moving into the area, they would move reserve troops to the front—something he wanted to avoid. He

Riva Ridge. (*Denver Public Library*).[c]

Riva Ridge tram. (*Denver Public Library*).[c]

1945 view of Mt. Belvedere. (*Denver Public Library*).

also wanted to avoid casualties and decided on a strategy of surprise. He moved four companies (a battalion and an extra company) to the base of steep Riva Ridge under cover of darkness. During the late night and early morning hours, the companies climbed up four single-file pathways, often having to drive steel pitons into the rock to keep going. At dawn, the 10th Mountain troops caught the enemy by surprise and took Riva Ridge with minimal casualties. The division's engineers, using steel cables, erected a simple aerial tram that was able to bring the wounded down off the ridge and resupply the troops on top in only a few minutes. Additional resupply, including pack howitzers, was done by mule train. Throughout the day and the next night, the soldiers on the ridge fought off seven counterattacks, but Riva Ridge remained in Allied hands.

During the night Riva Ridge was being scaled, the rest of the division marched eight miles to relatively concealed positions near the base of Mt. Belvedere for an attack the next night.

On the night of February 19, 1945, two regiments and a battalion from the division began the assault on Mt. Belvedere and its sister peak Mt. Gorgolesco. The soldiers were ordered to keep their rifle chambers empty and use only bayonets and hand grenades in the initial climb in the darkness. General Hays was worried the flash and sound of gunfire during the night would give away the attack too early, ruin the element of surprise, and give the Germans more time to counterattack. The 10th's engineers cleared pathways through the minefields, and barbed wire; and scout dogs helped guide the troops up to where the enemy was entrenched. Once the enemy line was reached, the 10th's soldiers were free to use all their weapons.

Part of General Hays' plan involved identifying and targeting all of the enemy artillery batteries in the area, as well as sixty-seven known mortar positions. The idea was that when the enemy started firing on our troops, our artillery would fire on the enemy artillery and mortar

Brazilians on Mt. Castello.[c]

positions. That is what Dad's battery did on February 20 and 21 as the battle wore on. The weather was favorable, and the Allies also used their air superiority to advantage, flying over four hundred sorties.

Tenth Mountain troops, under heavy opposition, took the top of Mt. Belvedere and attacked its sister peak to the east, Mt. Gorgolesco. By nightfall on February 22, the line from Riva Ridge to Mt. Gorgolesco was secure.

The whole offensive was called Operation Encore and involved not only the 10th Mountain Division but also the easternmost division in IV Corps, the Brazilians, who were to protect the 10th's right flank by taking Mt. Castello, an objective the Brazilians had failed to take on a number of previous occasions. With the 10th Mountain now on their left flank, the Brazilians this time were successful.

The final objective of the operation for the 10th Mountain was called Mt. della Torraccia, vital to control of Highway 64 and the Reno River Valley. The Germans hung on fiercely, and the 2nd Battalion of the 10th Mountain Division's 86th Infantry Regiment was decimated.

The 3rd Battalion—the one Dad's battery was supporting—was called in to relieve them. The 3rd Battalion also suffered heavy casualties as they fought their way up. Dad's battery fired about four hundred rounds on February 23 and the same number on February 24 in preparation for the 3rd Battalion's final advance to the summit, which was reached by about 7:30 p.m. There, the 3rd Battalion was told to dig in and await the eventual counterattack.

Mine sweeping on Mt. della Torraccia. (*Denver Public Library*).

On February 25, late at night, since Allied tactical air did not usually fly at night, the counterattack came. The Germans fired a barrage of a thousand artillery rounds against the dug-in 3rd Battalion before attacking; but when the German troops finally advanced, they were almost completely destroyed in a half-hour by deadly artillery from Dad's battery, and possibly others.

The six M7 Priests of Dad's battery fired 250 rounds on the counterattacking enemy between 11:30 p.m. and midnight—and the 3rd Battalion troops, tired and battle-weary as they were, hung on, never giving ground.

More of Highway 64 was now in Allied control. The test of the 10th Mountain Division's fighting ability was deemed an unqualified success.

Although nearly two hundred men in the 10th Mountain Division were killed and over seven hundred wounded in the offensive, the Mt. Belvedere-Mt. Gorgolesco-Mt. della Torraccia ridge was now securely in Allied hands, thanks to the courage of the infantrymen—and in part to a devastating new weapon.

19

THE VT FUZE

Artillery from Dad's battery and other batteries in the Fifth Army had become especially lethal due to what had been a top-secret weapon—first used on land in the European Theater at the Battle of the Bulge over the winter of 1944–45, and made generally available in the Mediterranean Theater at the end of January, 1945. Dad's battalion used it early in February on the coast in support of the 92nd Division. However, its devastating effectiveness against enemy personnel really made itself known in the 10th Mountain Division operation against the Mt. Belvedere-Gorgolesco-della Torraccia ridge. Before the US even entered the War, Congress agreed to help Britain by a program known as Lend-Lease (ultimately administered by Madison, Wisconsin businessman Leo T. Crowley). As part of the deal, knowing we would have a lot of catching up to do, Britain disclosed to the US a lot of secret weapons programs. The most famous was *radar*. Eventually, the Allied radar was far superior to anything the Germans or Japanese had. An offshoot of radar was an artillery fuze that emitted a radar signal which would cause a round to explode when it detected another object—or the ground. The Americans developed and improved on the concept and went into mass production.

First used by the US Navy, anti-aircraft rounds with the new fuze exploded when the fuze's radar pulse was within killing range of an enemy aircraft and proved devastating to Japanese planes trying to attack Allied ships. Some say the effectiveness of Navy anti-aircraft fire with the new fuze caused the Japanese in desperation to resort to low-flying kamikaze missions, where they faced devastating fire from Navy Bofors guns, whose smaller high-volume ammunition could not use the new fuze. The new radar fuze also was used in anti-aircraft artillery by the British and proved very effective against German V-1 rocket attacks in 1944—though useless against the later supersonic V-2s.

In field artillery, a high-explosive round with an ordinary *contact fuze,* sometimes called "superquick," exploded on impact, or, more correctly, a split-second later. When the projectile hit the ground, much of the explosive energy and shrapnel went into the ground. The most devastating kind of fire against enemy troops, whether in foxholes or in the open, was what was called an *air burst.* Prior to the development of the new fuze, an air burst was possible only with a mechanical *time fuze,* which required estimating precisely how long an artillery round would take to hit its target. With all of the variables involved, effective mechanical time-fuzed air bursts, at just the right height to do maximum damage, were notoriously difficult to achieve.

The new radar fuze, however—a proximity fuze, called a "VT" or *variable time* fuze, code-named POZIT—guaranteed an air burst at the correct height above the ground for maximum antipersonnel effect.

It was a new day in warfare. No longer was an enemy soldier safe in a foxhole or trench; and when in the open, not even "hitting the dirt" helped.

(VT fuzes, by the way, were also used for bombs. Sophisticated altimeter radar fuzes triggered the atomic bombs for air bursts at Hiroshima and Nagasaki).

The new VT fuze was at first used only over water for fear the Germans would recover one from a dud, reverse-engineer it, and use it against the Allies. As the War was winding down, however, the Allies calculated the War would end before the Germans could reverse-engineer it, and the VT fuze became available for antipersonnel use by Allied *field artillery* batteries, like Dad's.

Captured Germans in the Apennines told how they feared the inescapable Allied artillery, which had become even more lethal with the new fuze. The impact was not only physically devastating, but also psychologically demoralizing.

20

HARASSING FIRE AND DAD'S CLOSEST CALL

While the most important fire missions done by artillery batteries, like Dad's, were in support of infantry (or armor) troops in the field, most of the millions of artillery rounds fired by the Allies—and the enemy—in World War II were what was called "harassing fire." In my day, it was called "H&I" for harassment and interdiction. The stress of being under attack by shelling from distant artillery and mortars has long been known for its ability to demoralize an enemy and cause "battle fatigue" or "shell shock" over time. Therefore, artillery and mortars have always been primary weapons in psychological warfare.

High-explosive rounds were not the only kind used by the Allies in psychological warfare. Leaflets were sent to the other side in time-fuzed "smoke" rounds modified to release propaganda several hundred feet above the ground. The Germans did likewise.

While artillery and mortars were occasionally used for leafleting, their most common use in psychological warfare was for harassment,

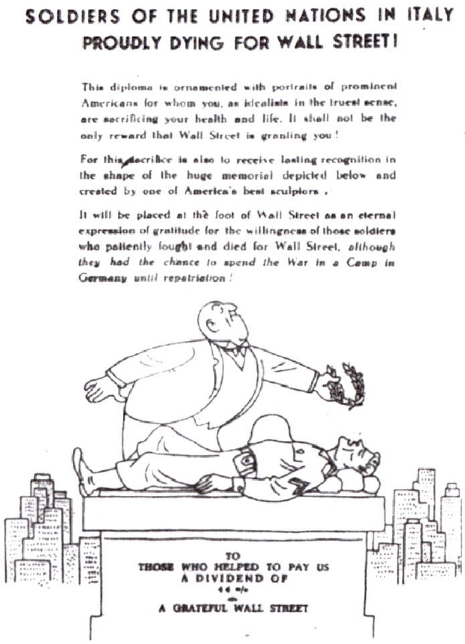
SOLDIERS OF THE UNITED NATIONS IN ITALY
PROUDLY DYING FOR WALL STREET!

This diploma is ornamented with portraits of prominent Americans for whom you, as idealists in the truest sense, are sacrificing your health and life. It shall not be the only reward that Wall Street is granting you!

For this sacrifice is also to receive lasting recognition in the shape of the huge memorial depicted below and created by one of America's best sculptors.

It will be placed at the foot of Wall Street as an eternal expression of gratitude for the willingness of those soldiers who patiently fought and died for Wall Street, *although they had the chance to spend the War in a Camp in Germany until repatriation!*

Passierschein

Der deutsche Soldat, der diesen Passierschein vorzeigt, benutzt ihn als Zeichen seines ehrlichen Willens, sich zu ergeben. Er ist zu entwaffnen. Er muß gut behandelt werden. Er hat Anspruch auf Verpflegung und, wenn nötig, ärztliche Behandlung. Er wird so bald wie möglich aus der Gefahrenzone entfernt.

OBERBEFEHLSHABER

Englische Übersetzung nachstehend. Sie dient als Anweisung an die alliierten Vorposten.

SAFE CONDUCT

The German soldier who carries this safe conduct is using it as a sign of his genuine wish to give himself up. He is to be disarmed, to be well looked after, to receive food and medical attention as required, and to be removed from the danger zone as soon as possible.

SUPREME COMMANDER,

German propaganda leaflet and Allied leaflet for safe passage.

usually at night. The idea was to interrupt sleep and curtail resupply and other activities unsafe during daylight when they could be spotted by aircraft or artillery forward observers. The harassing artillery, directed perhaps at a stretch of road, a building, a grove of trees—anywhere the enemy might conceivably be—was a constant reminder a war was going on, there was no safe time or place to relax—and death or serious injury could occur at any moment.

In the movies, we always hear the screech or whistle of the artillery round before it explodes. Dad told me that is not how it happens in real life. If you can hear the round, it usually is going safely well over your head. The round that hits you directly makes no sound until it hits. In other words, there is no warning. That had to weigh heavily on every soldier at the front.

I do not know where or when Dad's worst day—or night—in Italy happened. It might have been while he was on the coast in January and February of 1945. His battery experienced a lot of incoming artillery there. Or it may have been in the mountains. I only know the experience

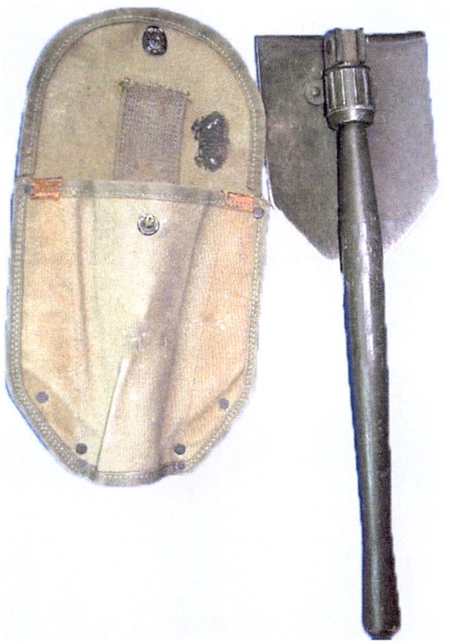

Standard-issue entrenching tool.

never left Dad. He relived it many, many times, throughout his life—as my sisters and I can attest.

Sleeping conditions for soldiers in a combat zone, like Italy, could vary a lot. I remember Dad saying for several weeks he even slept in an abandoned car. The northern Apennines were dotted with numerous rural towns and farmhouses with sheds, barns, stables, and other outbuildings. When such forms of shelter were available, the troops utilized them for sleeping. When out in the open, the soldiers used what were called "shelter halves." Two shelter halves put together made a two-man pup tent.

Almost all US soldiers were also issued a compact foldable shovel—called an "entrenching tool" in my day. It was another German invention, reverse-engineered by the Allies.

Of course, the soldiers also used regular-sized shovels and picks for digging, when they could.

Generally, when subject to incoming mortar and artillery fire, a soldier was safest in a hole in the ground called a foxhole. Some foxholes were dug as fighting positions and were made so a man could stand and shoot from them. Others were dug more for safety and for sleeping. On flat ground, the foxhole for sleeping had to be as long as a man and at least two feet deep. If available, a piece of wood, or something similar would be placed across the top of most of the foxhole, and dirt would be piled on top of it, leaving only enough room for the occupant to crawl in and out. Even better was digging into the side of a hill. As units moved and were replaced by other units, old foxholes often found new occupants. A foxhole was usually the safest place to be.

But not one night for Dad.

It happened when Dad was sound asleep in his foxhole, dug into the side of a hill. He was in his mummy sleeping bag, only his face exposed, in total darkness, dreaming of who knows what. He told me ordinarily, to avoid a condition called "trench foot," he slept barefoot with his socks on his chest in the sleeping bag, so he would have dry socks in the morning.

Suddenly a heavy mortar round, apparently harassing fire from the enemy, exploded very close to him. His whole body felt the shock wave as his foxhole collapsed—completely burying him. Dazed, in panic, his ears ringing, dirt in his mouth and eyes and nose, the weight of the earth on top of him—unable to move in his mummy bag—certain he had sustained possibly mortal wounds, utterly terrified—the most scared he had ever been in his life—he spit the dirt out of his mouth and screamed, "Medic! Medic! Medic!"

Two battery "aid men"—or "medics"—from the battalion's medical detachment were assigned to each firing battery. Artillery batteries, like Dad's, were usually on twenty-four-hour call. At least one man from the crew of each M7 Priest section had to be up to wake the others if a fire mission were ordered. The explosion of the mortar round in the darkness and Dad's cries for help soon resulted in several guys, besides the medics, coming to the rescue of the mess sergeant who had been buried alive. His fellow soldiers attended to badly shaken Dad. They dug him out of the collapsed foxhole and pulled him out of his sleeping bag. Though disoriented, and shaking, his hearing muffled, he was able to stand on his bare feet and had only minor cuts. I assume Dad, like all soldiers, had a first aid kit in a pouch that contained bandages and a powdered sulfa drug antibiotic. The medics administered first aid to Dad and offered to take him to the battalion aid station in their panel truck field ambulance.

If I have the story right, Dad was told, as someone receiving even only minor wounds under enemy fire, if he would go to see the doctor at the battalion aid station, he would be put in for a purple heart medal. Dad, however, still dazed, but grateful he was alive and not seriously wounded, and having seen men die and lose limbs for purple hearts, said he was okay and refused to go.

Dad was like that—but he later regretted not getting a purple heart for different reasons. More on that later.

Maybe his wounds were minor, but his fright and panic were not. The experience, while sound asleep, of a sudden explosion and shock wave, followed by being helplessly immobilized and buried alive, lived on in Dad's subconscious for the rest of his life.

When I was living at home, I don't know how many times I was studying late, or asleep, only to hear from my parents' bedroom Dad frantically yelling, "Medic! Medic! Medic!"

Sometimes an old World War II movie on television brought it on. Sometimes, it would happen when he was napping in his reclining chair.

No matter how often it happened, I was always alarmed to hear Dad's desperate calls for help and to see him afterwards in a cold sweat, having once again relived the terror of that moment—awakened from his nightmare by his own frightened screams.

21

TIME FOR A REST

On February 26, 1945, the day after the German counterattack on Mt. della Torraccia, Dad and a corporal and a private from Battery A were ordered to Fifth Army's rest camp in Florence for five days.

Fifth Army was concerned about morale, especially for the many troops who were spending their second winter away from home. Three rest camps—or "rest centers" as Fifth Army called them—had been established in Southern Italy and a large one in Rome. Late in 1944, two additional rest centers were established in Tuscany, one in Florence and one in a mountain spa town called Montecatini.

The idea of the rest center was to take the soldier out of the combat zone, let him sleep in a bed, take a hot shower, do some sightseeing and learn about the country, shop, enjoy bars and entertainment—in short, to feel human again, even if only for a few days. Although the US Army itself was segregated, multi-national Fifth Army was not, and discrimination on the basis of race or color was not tolerated in the rest centers, or elsewhere in the rear.

Pre-war photo of Florence's U-shaped, Fascist-era main railway station, still in use today, that became a Fifth Army "rest center" in late 1944-45.[c]

Ponte Vecchio in August 1944. Dating from 1345 A.D., it had no military value and was the only bridge left across the Arno River in Florence by the retreating Germans.

In a precision raid over art-laden Florence, Allied bombers had destroyed the rail yards when the city was still occupied by the Germans. The main railway station, however, known as Santa Maria Novella, received only light damage. So, after the Allies took Florence, Fifth Army converted the station into a giant rest center modeled on the one in Rome.

The Florence rest center had sleeping accommodations for 1,800 enlisted men and included showers, messes, barber shops, and motion pictures. Many of the sights of historic Florence, like the iconic Ponte

Vecchio—the only bridge across the Arno River left by the Germans—the Duomo, the Pitti Palace, the Palazzo Vecchio, Santa Croce, and others, were within walking distance of the rest center. The Germans had plundered many of the art works from the nearby Uffizi Gallery, which were not returned until later in the year, when the fighting was over.

The staff at the rest center was instructed to make the soldiers feel like they were civilians. The Army operated five restaurants at the rest center with the best food available, and beer and wine were free. There was also a large post exchange where Allied soldiers could buy cigarettes, toilet articles, candy, tobacco, beer, local souvenirs, books, magazines, and other items.

In addition, the American Red Cross operated a theater in Florence about two blocks from the rest center. A bath house and enlisted men's club were a few blocks farther away. The Red Cross was famous for its female volunteers, known as "donut dollies," who made hot coffee and donuts for the men in uniform. These American women were instructed to socialize with the men and make them feel at home. The WACs (Women's Army Corps) and Army Nurse Corps were also represented in Florence, I am sure.

The rest centers had a propaganda and informational purpose alongside the break from combat. Soldiers at the rest centers learned about other theaters in the world-wide conflict. They were shown some of the seven films in the "Why We Fight" series made in Hollywood and directed by Academy Award winner Frank Capra. The War Department also used the talents of Academy Award winning directors John Ford, John Huston, George Stevens and William Wyler to make films for the GIs about the War. For comic relief, the soldiers laughed at the irreverent Private Snafu cartoons made by Warner Bros. "Looney Tunes" department.

The movies were to entertain, but also to give the soldiers a sense of how their country appreciated them and a sense of their importance in the overall war effort where so many others, at home and abroad, were contributing. On an individual level, medical personnel gave talks on first-aid, hygiene, prevention of disease—and, of course, venereal disease.

What did Dad use for cash while in Florence? Dad, like most of the troops overseas, sent the major part of his pay home in the form of an

This 1,000 lire note in military currency was worth $10.

allotment. However, I am sure he received some of his pay in cash. Cash did not mean American money, or even Italian lire. Cash for an Allied soldier in Italy meant paper money—no coins—known as Allied military currency.

Called "AM-lire," or, commonly "Army lire," or "invasion money," the currency was worth 100 lire to the US dollar. The "funny money" was used to avoid inflation of the local currency (Italian lire) and to discourage black-market currency speculation. The soldiers were not supposed to have American money.

Rest centers like the one in Florence, encouraged athletic competition, such as basketball and baseball, as well as arts and crafts, for those so inclined. I am sure there were also pool tables and opportunities for card games. From what I recall Dad telling me, he was a fairly good poker player and sent his winnings home to add to the savings waiting for him there.

I have not found any pictures from Dad's brief stay at the rest center in Florence, though he was familiar with the city when we visited in March of 1994. In 1945 he must have been relieved to be away from the stress of the front and likely spent some of his time in Florence absorbing the atmosphere of the birthplace of the Renaissance, its history, its aromas, its numerous jewelry stores and leather shops, its famous churches, like the Duomo and Santa Croce, burial place of Michelangelo, Galileo, and Machiavelli.

I am sure, however, he did not appreciate Italian coffee—he only

Dad (lower right) having a gelato in Florence near the Ponte Vecchio. March, 1994.

liked American—or Italian food, which he never had a taste for. (I had a hard time getting him to try pizza when I was twelve or thirteen and was tired of the meat-and-potatoes fare we ate at home). The downside of being a mess sergeant was that Dad got to eat what he liked and never acquired the desire to try different foods.

I suppose at least once while enjoying the sights of Florence he must have thought to himself, "I sure am a long way from West Allis!"

But he was soon reminded of why he was in Italy. The fighting he had just left in the mountains was only the first phase of a two-phase offensive.

22

BACK INTO COMBAT

On March 2, 1945, Dad, and the corporal and private he left with, returned to Battery A from Florence. While he was away, the battery had moved forward from the spa town of Porretta Terme to near a small hamlet called Silla.

Fifth Army commander General Truscott, along with IV Corps commander General Crittenberger and 10th Mountain Division commander General Hays planned the advance west of Highway 64 in two phases so as not to alert the Germans a major offensive of division strength was going on. The generals did not want the Germans to send in a large number of reserves against the 10th Mountain Division. The second phase was supposed to start March 1, but bad flying weather delayed the start until March 3, after Dad had rejoined his unit.

With Allied air power having gained not only air superiority but air supremacy, Fifth Army worked closely with the Air Force in developing strategic bombing plans as well as tactics for close air support of ground operations. Dad was probably unaware of the interdiction bombing going on in the Po Valley. All the bridges over the Po River had been destroyed.

Railroads and highways used to move and resupply Axis troops were also bombed, as well as supply depots.

But Dad definitely saw the fighter-bombers going after the enemy troops in the Apennines. Close air support for Allied troops, like the 10th Mountain Division, was supplied by the XXII Tactical Air Command, based in Florence. Fifth Army had developed two tactics for using air power to support ground troops: "Rover Joe" and "Horsefly." Both, of course, required good flying weather.

Rover Joe was a forward ground observation team that included a pilot, usually rotated every fourteen days from a P-47 Thunderbolt or P-51 Mustang squadron. The pilot on the ground had a radio that enabled communication with incoming aircraft. He could reassign aircraft from fixed targets to targets of opportunity, like moving columns of tanks or trucks that had not appeared until after the planes took off. The Rover Joe pilot could also recommend angles of attack and give other information helpful to the pilots of the attacking fighter-bombers.

A "Rover Joe" unit, including a pilot, on Mt. Belvedere. (*Denver Public Library*).[c]

A "Horsefly" observation plane directing fighter bombers to targets.

Because he was on the ground, the Rover Joe pilot could not see some targets of opportunity that arose beyond his sight. For this reason, a second type of forward air controller was also employed, called "Horsefly." This was a pilot in an observation plane, usually a Stinson L-5 Sentinel. He could see the battle from above and even behind enemy lines.

In the event of a successful Allied infantry attack, the Germans would either retreat or bring in reinforcements. In either case, the enemy would be on the move instead of dug-in, and therefore would be vulnerable to attack from the air. Obviously, the Allies preferred attacking in good flying weather whenever possible, and the onset of a major attack was usually delayed if the weather was bad. Unlike the flyboys, troops in artillery units, like Dad's, operated day and night, in good weather and bad. But, as I mentioned before, observation aircraft, like the L-4 Grasshopper, even made artillery more effective when the flying weather was good.

So, March 3, 1945, the day after he got back from Florence, Dad was in the thick of it again. Right away, Battery A was on the move, along with the infantry units. The first day of Phase 2, Dad's battery moved twice, settling in just south of Mt. Belvedere at a place called Bombiana, which was 3–4 miles closer in range to the 10th Mountain troops that pushed off from locations in the heights this time. Dad's battalion had an observation post atop Mt. Belvedere, and the targets were on the other side of the mountain from Dad's battery.

The objective of Phase 2 was to dislodge the Germans farther from the high ground, free up more of Highway 64 and the Reno River Valley for the armor, and provide a launching pad for the 10th Mountain Division to spearhead a drive by the infantry over the hills down into the Po Valley, west of highway 64 (General Truscott's secret plan), when the Big Push came in April.

Leaving from Mt. della Torraccia, the peak that had come so dearly in Phase 1, the 10th Mountain Division troops attacked another ridge about four miles away including several small hamlets with a population of under one hundred and an important crossroads town called Castel d'Aiano. The dozen, or so, peaks that made up the ridge included Mt. della Spe, critical because it controlled Highway 623, the main German supply route from Modena in the Po Valley.

The taking of this high ground not only impeded the enemy's ability to resupply the Gothic line, but also, via Highway 623, opened up a range of potential attack routes on the northern slope of the Apennines leading down into the Po Valley, including the Panaro River Valley/Highway 4, about ten miles west of the Reno River Valley/Highway 64. It also put the 10th Mountain Division in position to take the town of Tolè, which led to the Samoggia Creek Valley/Highway 27 in between the Panaro and the Reno.

In fact, Tolè could have been taken as part of Phase 2, but the Germans started bringing in reserves, so General Truscott halted the advance on March 5, after only two days, not wanting to draw any more attention than necessary to his secret plan of attack for the spring offensive a month or so away. As part of Phase 2 the Brazilians, having proved to be a first-rate fighting unit, continued on the right flank of the 10th Mountain Division taking the town of Castelnuovo overlooking Highway 64 in the Reno River Valley.

In the face of Allied artillery and close air support, the Germans fiercely counterattacked. Elements of the 29th Panzer Grenadier Division, brought in from reserves, suffered heavy losses as the Allied troops held their ground.

Dad's battery fired over three hundred rounds on March 5–6 during four unsuccessful nighttime counterattacks by elements of the German 15th Panzer Grenadier Regiment against Mt. della Spe. Dad's battery fired harassing fire on March 7 and knocked out an enemy tank on March 8 as the front died down.

Losses to the 10th Mountain Division in Phase 2 amounted to more than two hundred killed and nearly eight hundred wounded. Over a thousand enemy soldiers were captured.

With additional miles of Highway 64 in the Reno Valley now open, tanks of the 1st Armored Division, in reserve most of the winter, took

Fifth Army commander General Truscott confers with IV Corps commander General Crittenberger (left) near Castel d'Aiano. (*Signal Corps*).[c]

up positions on the highway, while the 85th Infantry Division was put in reserve. Bologna, the prime objective in the Po Valley, as it had been for many months, was twenty miles away on Highway 64 and ten or twelve miles away on Highway 65.

Allied infantry and armor were both in good positions for the coming spring offensive.

On March 9, 1945, Dad's battalion relocated to near Riola, and Dad's battery set up in the vicinity of a hamlet called Collina di Savignano, east of Highway 64, where it was to remain for about a month.

German prisoners, guarded by 10th Mountain Division troops, carry a wounded comrade. (*Denver Public Library*).[c]

23

THE ROCCHETTA MATTEI

The success of the 10th Mountain Division and the Brazilians in the IV Corps sector during the limited offensive resulted in changes in responsibility for various units, both in IV Corps and in II Corps. In part, these changes were to prepare for the upcoming offensive and in part to allow units involved in recent fighting to spend time in rest areas so they would be fresh when it came time for the spring offensive in April.

They were also part of a deception.

With the Reno River roughly separating Fifth Army's IV Corps sector to the west from the II Corps sector, Dad's battalion was now technically in II Corps' area and was attached to the 34th Division Artillery. The 81st Cavalry Reconnaissance Squadron was detached from the 1st Armored Division and attached to the 10th Mountain Division for patrolling purposes. Dad's battalion established one forward observation post in Castelnuovo, west of the Reno River, and another at a place called Mt. Stanco, east of the river. Dad's battalion was given the mission of supporting the 81st Cavalry Reconnaissance Squadron on its

Photo of the Rocchetta Mattei castle. (*From Dad's negative. March/April 1945*).[c]

The Rocchetta Mattei today.

right and the 10th Mountain Division on its left as Fifth Army continued patrolling and intelligence-gathering in preparation for April.

Collina di Savignano, where Dad's battery stationed itself, was near one of the most iconic structures in the Reno Valley, if not in all the Northern Apennines, a place that looked like a fairy-tale Moorish castle. It was called *Rocchetta Mattei,* and Dad's battalion set up headquarters there. Dad left us a negative—among the few negatives I have found—so he apparently took a picture of it.

While it looks like a castle from medieval times, and was constructed on the foundation of an old castle, the Rocchetta Mattei was actually built in the mid-nineteenth century by a wealthy doctor from Bologna, Count Cesare Mattei, who invented a type of medicine he called *electro-homeopathy,* popular in the late nineteenth century, and today considered quackery.

Brig. General David L. Ruffner. Head of 10th Mountain Division Artillery, (*Denver Public Library*).

On or about March 16, 1945, Dad's battalion was attached to the 10th Mountain Division (where it remained until near the end of the war in Italy). The commander of 10th Mountain Division's artillery was Brigadier General David L. Ruffner, who worked well with General Hays, commanding general of the 10th Mountain Division, and a former artillery officer himself. General Ruffner would basically be deciding where Dad's battalion would be locating and what 10th Mountain Division units Dad's battalion would be supporting during the spring offensive everyone was anticipating.

After Dad's battalion was attached to the 10th Mountain Division, it moved its headquarters from the Rocchetta Mattei castle to a place called Palazzo, along with Batteries B, C, and the attached Battery X (751st Tank Destroyer), to support troops of the 86th Infantry Regiment of the 10th Mountain Division.

As the M7 Priest was a *howitzer* mounted on a lightly-armored tank chassis, so a "tank destroyer" was an anti-tank *gun* mounted on a lightly-armored tank chassis. American tank destroyers usually had turrets, like a tank, but the turrets were open on top. As the war went on and fewer German tanks were deployed, the Allies made other use of their tank destroyers, which could fire high-explosive as well as anti-tank rounds, had a greater firing range than the howitzers, and were particularly effective in knocking out sniper nests in urban areas. During this time, Dad's battery, Battery A, remained in the vicinity of La Collina di Savignano and the Rocchetta Mattei in support of the 81st Cavalry Reconnaissance Squadron, which had been attached to the 10th Mountain Division.

The front, however, remained relatively quiet.

24

MONTECATINI

On March 26, as part of Fifth Army's preparation for the April offensive, Dad's battery was relieved from combat duty and taken by motor convoy to the Montecatini rest center in Tuscany in the mountains between Florence and Lucca. I am not sure Dad was with them.

Montecatini, or Montecatini Terme, by its full name, had been known for its spa waters since Roman times and became a popular resort in the eighteenth and nineteenth centuries—the three oldest of the spas dating from the 1770s. Fifth Army basically took over the city to accommodate up to an entire division (10,000–15,000 men) at a time. All available hotels, boarding houses, and other lodgings, more than three hundred in all, were requisitioned. Montecatini's sulky horse race-track had

American Red Cross service club in Montecatini. (*Signal Corps*).[c]

As with most of Dad's photos, this one is undated and gives no location. Since he and his buddy on the right have fresh uniforms, it might have been taken after a uniform exchange.[c]

ceased operating due to the war, and two hundred winterized pyramid tents were set up there to handle additional troops. If they wanted, the troops could choose from among five different types of mineral waters, luxuriate in mineral baths, and have massages. The food was fresh and good. Ice cream was served daily. PX and souvenir shopping was available, as in Florence. The American Red Cross operated two service clubs.

Montecatini was a city of wide streets and beautiful gardens. It was a favorite of composer Giuseppi Verdi, and many of the buildings dated from the Belle Epoque of the late nineteenth century. Fifth Army opened four movie theaters, and various USO troupes offered daily live entertainment. Old blankets could be turned in for new ones, as could old uniforms.

The purpose of the rest center was to get the soldiers away from the front with clean beds to sleep on and otherwise to make them feel almost like civilians.

On April 1, Easter Sunday, 1945, after four days at the rest center in Montecatini, Battery A was taken by motor convoy back to their M7 Priests at La Collina di Savignano.

Soldiers enjoying shoe shine (or boot shine) at Montecatini.(*Signal Corps*).

The same day, they had a fire mission and fired thirty rounds on enemy personnel.

Strategically, on Easter Monday, April 2, 1945, after a month-long battle, the Soviets succeeded in capturing the Axis oilfields at Nagykanizsa, Hungary—the last significant source of petroleum for the Nazis.

The vaunted Nazi war machine was now almost entirely dependent on synthetic fuel made from coal in plants that were under the constant threat of Allied bombers.

25

LA COLLINA DI SAVIGNANO

The few negatives Dad left us, some in very bad shape, apparently are from one roll of film. I believe they are of Dad's battery area at La Collina di Savignano, because they contain the Rocchetta Mattei nearby. Somehow, I have a vague recollection Dad said he bought a camera and lost it right away. He may have bought a camera in Montecatini, or maybe my memory is faulty. He is not in any of the photos, so I assume he took them. They give an idea of what life was like in Dad's unit, Battery A of the 1125th Armored Field Artillery Battalion, during a quiet time on the front in a relatively safe area, most likely late March or early April of 1945.

The men in Dad's battery had to be ready for fire missions 24/7. They took turns on guard duty at night. At least one man from each gun crew had to be awake in the event of an emergency. They also took turns visiting Dad's kitchen truck or mess tent to eat. When the infantry was not active, there was a lot of down time for the artillerymen. They were often bored. As you can imagine, rumors were rampant. The soldiers were always

Dad's battery was set up in a farmer's field. Note the conical hay bale, typical of the Apennines. The pyramid tent on the right might have been a mess tent. When relocating, each M7 Priest towed an M10 ammunition trailer, like in left center of photo. The Army's workhorse trailer, however, was the one-ton G-518 "Ben-Hur," named for the Ben-Hur Mfg. Co. of Milwaukee. (*From Dad's negative*).[c]

Ammunition for the M7Priest was kept nearby.(*From Dad's negative*).[c]

The men could not venture too far, in case they were needed for a fire mission. (*From Dad's negative*).[c]

Beer was a special treat. (*From Dad's negative*).[c]

These guys slept on cots. The Time magazine in the foreground is from March 5, 1945. (*From Dad's negative*).[c]

Ha! Typical Dad. This soldier thought he would have privacy on a portable field latrine. (*From Dad's negative*).[c]

hungry for information. They talked with each other, played cards and games, and wrote letters.

Mail call was always a special time. To handle mail delivery between the front and the States, the War Department and the Post Office adopted a process used by the Brits, a type of letter called "V-Mail," or Victory Mail, to save the weight and bulk of transporting paper letters. A "V-Mail," used by GIs overseas and civilians at home, was a one-page letter written on a standard two-sided 8 1/2" x 11" form that could be folded and sealed—no envelope required.

Two guys in Dad's battery writing home. (*From Dad's negative*).[c]

A fellow soldier, showing another view of Dad's battery area. (*From Dad's negative*).[c]

After mailing, the V-Mail letters were opened and photographed on 16mm microfilm. The film was flown overseas or back to the States, as the case might be, where photographic machines printed paper copies from the film. It was said one mail bag of microfilm could hold the equivalent of 136,000 letters, saving precious weight and room in transport planes. After clearing and stamping by wartime censors, the 4" x 5" paper copy printed from the microfilm was put in a window envelope—with the addressee's address showing—and delivered. The whole process took only a few days compared to bulk mail that went by ship and took weeks. I am sure Dad, like all GIs, loved getting mail. Unfortunately, no V-Mail or other correspondence to or from him during his service has survived, as far as I know.

One amusement for men, like Dad, on the front, was radio. Fifth Army actually had its own mobile radio station that featured news and entertainment for the troops.

The station was actually two trailers, one with a small studio and the other with the transmitter. Most of the time Dad was in the mountains, the station, which had a range of about fifty miles, was based in Florence. Calling itself the Fifth Army Mobile Expeditionary Radio Station, it played transcribed music and radio programs from the Armed Forces Radio Service, based in Los Angeles, as well as live news, interviews, and even live dramas. An outdoor studio was able to accommodate an Army band and visiting USO entertainers. Though aimed at American GIs, the station was also popular with local Italians and even enemy soldiers who liked American music.

Fifth Army's radio station had competition, however, from a Fascist Italian station broadcasting from Milan that featured an American woman with a sultry voice who called herself "Sally" like the original "Axis Sally" in Berlin. She did her best to make the American troops homesick and tried to lower their morale. Usually, the men just enjoyed the American music she played and laughed at Sally, whose real name was Rita Luisa Zucca, daughter of a Manhattan restaurant owner.

I am sure, while on the front, Dad listened to Fifth Army's radio programming, as well as to the Italian Sally. The BBC also had an Italian broadcast service that was popular with the GIs.

26

MORE REST?

According to a morning report, on April 3, 1945, only two days after his battery returned to La Collina di Savignano from the rest center at Montecatini, Dad and two other enlisted men were relieved from duty to go back. In addition, four more enlisted men were relieved from duty to go to the rest center at Florence. I do not recall Dad ever telling me about Montecatini, much less that he had been there twice. The morning reports from that time show Dad would have been responsible for feeding eighty-eight men, yet the photos we have of the battery area from Dad's negatives show very few soldiers. Speculation on my part, but it is possible the photos from Dad's negatives were taken while most of Dad's battery (i.e., eighty-eight men) were in Montecatini and he and a few others were left behind as a skeleton crew to guard the equipment—and then Dad and the others got to go on leave once the rest of the battery returned. Hence, we cannot be sure Dad went to Montecatini twice. He may have gone only once, starting April 3.

In any event, he was back on duty by April 7, 1945—in time for his battalion's involvement in the Big Push.

27

GERMAN CHANGE OF COMMAND

In March of 1945, the war was going badly for Germany, not only in Italy but also on the Eastern and Western Fronts. The architect of the German defense in Italy was Field Marshal Albert Kesselring. He brilliantly fought the Allied advances from the time the Allies first invaded the Italian mainland in September of 1943, making a series of enlightened strategic retreats. A favorite of Hitler's, Kesselring was totally loyal to the *Führer*. In October of 1944, Kesselring was seriously injured in a motor vehicle accident, and General Heinrich von Vietinghoff was temporarily put in charge of the German armies in Italy for a few months at the end of 1944 while Kesselring recuperated.

On the Western Front in early March of 1945, the Allies approached the Rhine River in Germany. Hitler, wanting to make a stand at the river, ordered all bridges on the Rhine destroyed. When he learned the Allies made it to one bridge, at Remagen, Germany, south of Bonn, and crossed it before it could be completely destroyed, he was furious. He

Field Marshal Albert Kesselring.

General Heinrich von Vietinghoff. (*Bundesarchiv Bilder*).[c]

ordered the deaths of the senior officers with responsibility for destroying the bridge and relieved Field Marshal Gerd von Rundstedt, who had been in charge of the Western Front and who had recommended to Hitler unconditional armistice on grounds the situation was hopeless.

On March 10, 1945, or so, Hitler removed Field Marshal Kesselring from Italy to replace von Rundstedt and put him in charge of defending Germany on the Western Front. General von Vietinghoff was named to replace Kesselring as commander of the German armies in Italy, permanently this time. Kesselring had favored a policy of conserving his troops by voluntary withdrawals to stronger defensive positions using delaying tactics, as had worked so well in Italy before; but Hitler, concerned with troop morale, forbade any more voluntary withdrawals in Italy. Retreat was to come only after enemy engagement—as it turned out, a disastrous policy.

Meanwhile, unknown to ordinary soldiers, like Dad, secret surrender negotiations were going on in Switzerland between American Allen Dulles of the Office of Strategic Services—forerunner of the Central Intelligence Agency—and Waffen SS chief Karl Wolff, head of the SS in Italy and Western Austria. The Waffen SS was the military wing of the

General Karl Wolff, head of SS in Italy. (*Bundesarchiv Bild*).[5]

Nazi party, and its units fought beside the regular army (part of the Wehrmacht) troops in Italy.

Wolff had a clearer idea than the army of troop morale and the Italian public. He feared the coming Allied offensive in Italy would be a futile bloodbath and decided to try to convince Kesselring to surrender in Italy. Kesselring's reassignment, however, frustrated Wolff's plans. Wolff knew Kesselring quite well—not so von Vietinghoff. Wolff also knew his bargaining position with the Allies, minimal as it was in light of the unconditional surrender demand, would diminish once the Allied offensive began.

The Soviets, apprised of Wolff's overtures, feared such negotiations would amount to a "stab in the back" and free up the German divisions in Italy to fight them on the Eastern Front. Stalin insisted a Russian be present—which would have stopped the talks cold—and complained to the President.

Dulles was ordered to break off negotiations, but, as we shall see, the contacts continued informally, on-again, off-again, through the Swiss until the fighting in Italy ended.

Wolff told the Allies the German strategy was to hold the Gothic line as long as possible. If and when necessary, the Germans intended to retreat out of the Apennines to the Po Valley and make another stand at the Po River. The next stand was expected to be taken in the Alps above the Adige River, a tributary of the Po. If necessary, a final retreat would come behind the Adige line defenses through the Brenner Pass in the Alps between Italy and Austria. Wolff told Dulles of an incredible final plan. Hitler and the Nazis intended to make a last stand at a redoubt, or fortress, in the Bavarian or Austrian Alps where a doomsday secret weapon would be unleashed on the Allied armies.

Wolff, who hardly knew von Vietinghoff, painted a picture of the new commander in Italy as an old-line general who took seriously the loyalty oath sworn by all German soldiers:

> *I swear to God this sacred oath that to the Leader of the German Reich and people, Adolf Hitler, supreme commander of the armed forces, I shall render unconditional obedience, and that as a brave soldier I shall at all times be prepared to give my life for this oath.*

Von Vietinghoff was aware of the shame suffered by German troops in the surrender that ended World War I and would not tolerate his brave troops being humiliated. According to Wolff, General von Vietinghoff was aware of the hopelessness of the Nazi cause and the hopelessness of the Axis military situation in Italy, especially in light of Allied air power. He knew of the good treatment afforded to captured German soldiers by the Allies, particularly the Americans, and despised and feared the newly-armed Italian partisans whose country the Germans were supposedly defending, and who did not observe the rules of warfare.

Ultimately, the reasonable approach to avoid further bloodshed, urged by Karl Wolff to von Vietinghoff, was rejected, temporarily, at least. Von Vietinghoff, despite the likelihood of defeat, was nevertheless prepared to sacrifice many of his soldiers and preferred honorable defeat to surrender, which he saw as disloyalty, at least as long as he had under his command armies that could fight.

. . .

Dad told me that the Allied soldiers, near the end of the war, feared the Germans, in desperation, might resort to chemical warfare—poison gas—as had killed and injured so many in World War I. Fortunately, Dad never had to use his gas mask, which would not have worked anyway against deadly nerve agents the Germans had recently invented and produced in quantity— and for which the Allies really had no answer.

I recall Dad telling me a lot of the World War I veterans, who had been gassed, suffered lung damage and died young during the 1930s. A lot of the gassed World War I vets also were known to have come back home sterile, unable to father children.

Most experts on World War II cannot explain why the Germans, abiding by the Geneva Protocols—agreed to between the World Wars—fought to the end entirely with conventional weapons. Perhaps Hitler viewed nerve agents and poison gas, in general, as too difficult to control—or perhaps he feared the threat of reprisals from the Allies, who also had chemical weapons in-theater.

On the evening of December 2, 1943, the worst chemical attack of World War II occurred when 105 Junkers Ju 88 Luftwaffe medium bombers mounted a surprise raid on the Italian port of Bari on the Adriatic—sinking or destroying twenty-eight Allied cargo vessels with a loss of only two aircraft. A thousand Allied military men and merchant marines were killed. The disaster did not end when the bombers left. One destroyed US Liberty ship, the S.S. John Harvey, had a secret cargo of mustard gas bombs to be used if the Germans resorted to poison gas. The bombed vessel caught fire and exploded, spewing a deadly cloud of mustard gas over the port and onto the oil-slick water, where survivors of the raid were fighting for their lives. Over six hundred military personnel were hospitalized with burns, blindness, and lung damage due to the mustard agent, and over eighty subsequently died. Civilian casualties near the port were estimated to have been even higher. (For decades, Allied authorities covered up that many of the raid's casualties were caused by mustard gas from one of their own cargo ships).

Alternatively, Hitler's reasons for not using poison gas may have been personal. A British mustard gas attack in World War I had blinded him for many months in 1918. Whatever his reasons, despite all his crimes, and his use of gas in extermination camps, Hitler, even at the end, forbade use of poison gas in battle—much to the relief of ordinary Allied soldiers, like Dad.

28

PRELIMINARIES

While Dad was still in Montecatini eating ice cream and donuts, going to movies, and enjoying himself at the Fifth Army rest center there, the first stages of the *Big Push* began.

Before attempting a knockout punch toward Bologna in the Po Valley behind the Gothic Line, the two armies of 15th Army Group, the American-led Fifth Army in the west and British-led Eighth Army in the east, decided to throw a few feints and jabs. The primary purpose was to try to draw the German reserve troops away from the center, where the main thrust of the Allied attack was to occur, or to keep enemy reserve troops from going there.

"Big Cover" was the name of a hoax the Allies succeeded in pulling off. The idea was to make the Germans think the 85th and 88th Divisions were moving from Fifth Army's II Corps east to be attached to Eighth Army, thinning out the Allied forces in the center; and IV Corps was supposed to look like it was spreading east and taking over II Corp's area of the front All the while, of course, the plan was to have the main attack out of the mountains along parallel highways 64 and 65 in the II

Corps zone leading to Bologna in the Po Valley. Fictional command posts were established, supply dumps camouflaged, false radio transmissions given, and artillery movements undertaken to effectuate the ruse. (The move of Dad's battalion to La Collina di Savignano east of the Reno River and Highway 64 in the II Corps zone likely was a part of this deception).

The Big Push (Operation Grapeshot) was to start with Fifth Army movement along the west or Ligurian Coast, followed by a thrust by Eighth Army (falsely looking like it had been beefed up by Fifth Army's II Corps) along the east or Adriatic Coast. Then the 10th Mountain Division in Fifth Army's IV Corps (the division Dad's battalion was attached to) was to advance north in the mountains west of the Reno River Valley/Highway 64, until it was about even with II Corps, at which time IV Corps and II Corps would advance together toward the Po Valley, with II Corps taking Highways 64 and 65, and the hills in between, toward Bologna, roughly through the middle of the Gothic Line. The Germans, in other words, had to try to figure out if the main Po Valley attack was coming on their right flank, their left flank, or still out of the mountains and down the middle toward Bologna, where their heaviest defenses were concentrated.

In terms of strength, Fifth Army had a total of nine divisions and Eighth Army eleven, plus various independent brigades. They were facing an Axis force in northern Italy consisting of twenty-four under-strength German and five Italian Fascist divisions.

Fifth Army had about 270,000 troops, plus another 30,000 in replacement depots ready to substitute for the killed and wounded. It had about 2,000 artillery pieces of one kind or another, more than twice the number of the entrenched Germans. Captured German prisoners were frequently cooperative in providing information about enemy troop strength and morale. As expected, German morale was generally low. However, certain hard-core units that had fought together since the Afrika Corps days seemed willing to fight to the end. Many of the younger German recruits were more afraid of being shot in the back for desertion than they were of the Allies.

The spring offensive began in the western coastal area where Dad had spent January and part of February. The formerly all-African-American 92nd Infantry Division was now comprised of one African-American regiment (the 370th), one Japanese-American regiment (the 442nd Regimental Combat Team—Nisei), and one Caucasian regiment (the 473rd, former anti-aircraft artillerymen).

Dad, I think, was always thankful he ended up in field artillery and not the infantry when his anti-aircraft unit was no longer needed. I remember when I went into the Army in 1969, he told me to avoid the infantry, with its hardships and danger, at all costs.

General Almond, commanding general of the 92nd Infantry Division, had much the same plan in April as was tried so unsuccessfully in February. Starting April 5, the 442nd Nisei would attack in the hills and the 370th (African-American) would advance along the coast, but closer to a highway in the foothills—rather than the beach, as in February. The 473rd (Caucasian) regiment was to cover the flank of the Nisei. The 442nd Nisei—which became the most highly decorated unit of its size in World War II—achieved in about a day the objective that was not accomplished in February. It moved quickly north taking the high ground, and the 370th could not keep up. General Almond and General Truscott then had the 473rd and the 370th basically switch places. By April 7, the 442nd in the hills had outflanked Massa on the coast; and by April 9, the 473rd had fought through mine fields and heavy resistance in the coastal area to surround the town. On April 10, Massa was finally taken. The next objective was Carrara (famous for its marble), and then on to the Italian Fascist naval base at La Spezia.

The Italian long-range coastal guns guarding La Spezia at Punta Bianca continued to be a problem for the advancing 92nd Division, as they had been in February. At the beginning of the 92nd's offensive in April, Allied warships pounded the coastal artillery battery from afar, but to no avail. Likewise, almost daily bombings by Allied fighter-bombers failed to silence the guns which were most effective against Massa and Carrara. When they were finally within range of Allied field artillery, the coastal guns became the target of all thirty-six 76mm guns of the all-African-American 679th Tank Destroyer Battalion, which eventually fired over 11,000 rounds on Punta Bianca and silenced the enemy coastal guns by April 19.

Although the 92nd Division's coastal advance was halted somewhat by about April 14, it had succeeded in putting Fifth Army in a good position to continue up the coast and had caused the Germans to divert a battalion of reserves all the way from Modena, near Bologna, where Fifth Army's main attack was coming.

. . .

On the *Western Front*, by mid-April of 1945, about the time of the Big Push in Italy, and into May, American and British troops began encountering and liberating Nazi concentration and death camps in Germany and elsewhere. Intent on murdering Jews, Romani, and others on an industrial scale, the Nazis had built extermination camps outside of Germany, mostly in Poland—to hide their embarrassing existence from the German people. The Russians freed the first of seven such camps (Majdanek) as early as July of 1944—so Allied intelligence knew of their existence by then. In November of 1944, with the Soviets advancing on the Eastern Front, the Nazis began dismantling the gas chambers and crematoria and tried to move as many of the starving slave-labor, death-camp inmates as possible west—to concentration camps in Germany.

Dad told me he and the other ordinary GIs in Italy were not aware of death camps, and had no idea the Nazi regime was engaged in the mass murder of the Holocaust—until the gruesome films of it appeared, after the fighting in Italy was over.

Of the forty to fifty thousand Jews who lived in Italy before the War, an estimated eight to ten thousand were killed in the Holocaust. Most of those who perished were captured during the *Nazi occupation* of Italy and transported by rail to death camps in Poland—mainly Auschwitz-Birkenau. Although the Italian Fascists did not establish death camps, the Nazis, after they invaded Italy in 1943, converted the buildings of the San Sabba rice-husking mill in the Adriatic port of Trieste, Italy, into a notorious prison camp. While the camp was primarily used as a temporary detention center for those in transit to death camps in Poland, a crematorium was eventually built at San Sabba, and an estimated three to five thousand Italian and Yugoslav partisans and political prisoners, Jews, and other inmates were executed there.

• • •

After the Nazi Holocaust came to light, and in the spirit of *never again,* the United Nations in late 1948 adopted an international treaty called the *Convention on the Prevention and Punishment of the Crime of Genocide.* The United States, under President Harry Truman, was the first nation to sign. However, as a treaty, the pact would not be binding on the US until Senate ratification; and the Senate, fearing loss of sovereignty, and with negative input from the American Bar Association, tabled ratification.

Years went by with little indication the *Convention* would ever receive Senate ratification.

Then, in January of 1967, Wisconsin's maverick Democratic Senator William Proxmire took up the cause of the long-pending Genocide Treaty, calling the failure to ratify it a "national shame." Senator Proxmire, known for his sterling attendance record, doggedly vowed to deliver a speech *every day* the Senate was in session to remind his fellow senators of their failure to ratify the 1948 UN treaty and their duty to do so.

After *nineteen more years*—and over three thousand daily speeches—Senator Proxmire finally succeeded in getting the Senate to ratify the Genocide Treaty, which then required Congress to pass a law implementing the terms of the treaty.

Finally, on November 4, 1988, some forty *years* after the UN had adopted the Genocide Treaty, President Ronald Reagan signed into law the *Genocide Convention Implementation Act of 1987* (known as the **Proxmire Act**), which made genocide a federal crime and made the US a full partner in UN efforts to ban genocide throughout the world.

29

THE BIG PUSH

On April 7, 1945, Dad returned to Battery A at La Collina di Savignano from the rest center at Montecatini. His battery fired on enemy personnel the same day. On April 8, the battery fired on an enemy strong point. On April 9, Dad's battery fired on an enemy anti-tank gun and three hundred rounds against enemy personnel. The same day, while Dad's battery remained at La Collina di Savignano, battalion headquarters and the other two M7 firing batteries in Dad's battalion moved to a place near Sprille.

Also on April 9, the British-led Eighth Army began its devastating attack on the eastern end of the Gothic Line near the Adriatic Coast against a number of rivers, lakes, flooded areas, and other water barriers. The plan was for Eighth Army to make an end run around the Gothic Line in the east, and II Corps of Fifth Army to break through the center of the line in the mountains, the two forces then surrounding Bologna in a pincer movement.

To overcome the Germans' use of their very efficient mechanized reserves, the Allies intended to attack, as much as possible, along the entire front rather than try to punch through the enemy line in one place. For Fifth Army, the plan called for IV Corps and the 10th Mountain Division (and Dad) to start the main offensive in the mountains. When the 10th came abreast of II Corps, south of Bologna, protecting II Corps' left flank, both IV Corps and II Corps would come out of the mountains together, as much as possible. II Corps was to surround and take Bologna while IV Corps and its 10th Mountain Division would remain west of Bologna and continue to drive north to prevent the Germans from forming another defensive line at the Po River. If possible, the offensive would continue all the way north across the Po Valley to the Adige River in the Alps, near Verona, where construction on a formidable new German defensive line had been going on for nine months.

On the evening of April 10, Dad's battery was relieved of its support of the 81st Cavalry Reconnaissance Squadron. The next day, Dad apparently went to Porretta Terme, an hour away, to receive a Good Conduct Medal, while his battery moved into position near a place called Borra, just west of Batteries B and C. The same day, Lieutenant Colonel Maurice Shaver, the battalion commander, briefed his officers. The battalion would be in support of the 2nd battalion of the 87th Infantry Regiment of the 10th Mountain Division, which would begin their attack in a few days. That night, German artillery harassing fire raked over the valley where Dad's battalion was located, wounding three men.

Field Marshal Alexander.

The attack was set for the next day, April 12, 1945. Dad probably knew he was about to embark on the most exciting time of his life. Like all the soldiers, he received a flyer, which he kept as a souvenir, from the top commander in the Mediterranean, Field Marshal Harold Alexander, wishing "Godspeed and good luck" to all.

Though Dad and the other Fifth Army soldiers were ready to push off on April 12, bad flying

ALLIED FORCE HEADQUARTERS

April, 1945

SPECIAL ORDER OF THE DAY

Soldiers, Sailors and Airmen of the Allied Forces in the Mediterranean Theatre

Final victory is near. The German Forces are now very groggy and only need one mighty punch to knock them out for good. The moment has now come for us to take the field for the last battle which will end the war in Europe. You know what our comrades in the West and in the East are doing on the battlefields. It is now our turn to play our decisive part. It will not be a walk-over; a mortally wounded beast can still be very dangerous. You must be prepared for a hard and bitter fight; but the end is quite certain — there is not the slightest shadow of doubt about that. You, who have won every battle you have fought, are going to win this last one.

Forward then into battle with confidence, faith and determination to see it through to the end. Godspeed and good luck to you all.

H.R. Alexander

Field-Marshal,
Supreme Allied Commander,
Mediterranean Theatre.

Ready to start the Big Push, Dad, like all the Allied soldiers, received this flyer. Dad's proud comment in the lower left: "We were delayed two days [by weather] after receiving this, but we carried it out in every detail."

weather grounded the attack aircraft and caused a postponement. All was on hold.

On April 13, with the planes still grounded, radio reports were received on the front that sixty-three-year-old President Franklin D. Roosevelt had died unexpectedly of a stroke at his retreat at Warm Springs, Georgia, the previous day, April 12, 1945, while sitting for a portrait. Harry S. Truman, a former World War I artillery officer, having been vice president only eleven weeks, was sworn in as president on April 13.

I remember Dad telling me he did not learn until Roosevelt's death, or perhaps shortly before, that the never-photographed-in-a-wheelchair four-term president had been paralyzed in 1921 by polio—some experts today suspect it was really Guillain-Barré syndrome—and could not stand without leg braces. [On March 1, 1945, about six weeks before he died, FDR apologized for sitting while addressing a joint session of Congress and acknowledged publicly, for the first time, he normally carried "about ten pounds of steel" around his lower legs.]

Roosevelt would not be the only wartime leader to die in April of 1945.

At 7 a.m. on April 14, the airfields were still shrouded in mist and fog. But within an hour, the clouds showed signs of clearing. By 8:30 a.m., Dad must have welcomed the sight of wave after wave of fighter-bombers overhead destroying the numerous targets identified by patrolling and intelligence.

Next came the artillery. Starting about 9:00 a.m., Dad's battery alone fired over a thousand rounds into enemy positions and against enemy personnel before, and during, the 10th Mountain Division's initial advance from Castel D'Aiano and Mt. della Spe, objectives taken in Phase 2 of the operation the previous month.

Unfortunately, the Germans were ready. As the various units of the 10th Mountain Division reached their respective departure lines, the Germans came out of their dug-in positions. The American infantrymen, as they pushed off, were at a disadvantage. To get to the enemy on the next ridge they had to cross a natural basin called the *Pra del Bianco,* where their actions were under constant enemy surveillance. They faced heavy artillery, mortar, machine-gun, and sniper fire as they crossed minefields to reach their objectives. Fighting that first day was intense, and casualties were high.

The night of that first day, a young, replacement second-lieutenant from Kansas in Company I of the 3rd Battalion of the 85th Infantry Regiment of the 10th Mountain Division was severely wounded in the back and arm while assaulting a German machine-gun emplacement. His name was Robert Dole. He later became US senator from the state of Kansas and was the unsuccessful Republican candidate for president in 1996, the last World War II veteran to be a nominee of a major party.

Dad's battalion also had casualties. Private Homer Stoughton of Headquarters battery, as part of a forward-observer team supporting Company F of the 2nd Battalion of the 87th Mountain Infantry Regiment, was killed by

10th Mountain Division troops approach Rocca di Roffeno. Note smoke from artillery fire. (*Denver Public Library*).[c]

enemy artillery fire—the first man in Dad's battalion to die in action. A second member of the same forward-observer team was also seriously injured. A new forward-observer crew from Dad's battalion went out immediately to support the infantry.

Progress was slow, but the 10th Mountain Division had a sense of mission and not only gained ground but also gained valuable intelligence doing things only the infantrymen could do. They learned the German division they were facing had committed its reserve regiment. They also learned an entire German division, the 90th Panzergrenadier, was in reserve in the event of a break in the Gothic line. By overrunning an enemy artillery observation post and capturing maps and a radio, they learned where enemy mortars and artillery were located. On the body of an SS sergeant major, they learned the German passwords for the next three days.

As the 10th Mountain Division was crossing the Pra del Bianco basin through the German line in the mountains and toward the crossroads town of Tolè, the Brazilians covered their left flank. On their right, the tanks of the 1st Armored Division attacked the heavily defended town of Vergato down in the Reno River Valley along Highway 64.

The next day, April 15, the six M7s in Dad's battery fired over two thousand rounds in support of the infantry's advance. But by that evening, the 10th Mountain Division's 87th Regiment was close to getting out of range of the protection of Dad's battalion. The same day, more than eight hundred heavy (i.e. four-engine) and 250 medium bombers announced to the enemy that II Corps was beginning its advance east of IV Corps between Highways 64 and 65. At dawn the next day, after the dust had settled, the South African 6th Armoured Division and the

88th Infantry sprang into action in II Corps between the two highways that led to Bologna. The 34th Infantry and the 91st also took objectives in the heavily defended II Corps sector where Fifth Army's advance had been halted short of Bologna the previous October.

On April 16, Dad's battalion was to move forward in the IV Corps area via the narrow, treacherous, winding mountain roads with their frequent switchbacks. At one steep, downhill hairpin turn, one of the M7 Priests of Battery C toppled off the road and landed in front of a half-track on the roadway below, killing one man and injuring seven others. The upended M7 blocked the narrow roadway for several hours. According to the aforementioned memoir of Leonard Warner, the battalion actually had two M7s overturn trying to negotiate up and down the narrow, twisting mountain roads.

Another casualty was a kitchen truck, like Dad's, apparently also from C Battery. The kitchen truck was pulling a full water-buffalo water trailer uphill when the truck stalled. As the driver clutched to downshift to a lower gear, the heavy trailer jackknifed, the truck's wheels spun in the dirt, unable to get traction, and the water trailer pulled the kitchen truck right off the road and down a cliff. Fortunately, the driver was able to jump out and avoid injury.

Meanwhile, Battery A, Dad's battery, was ordered to take an alternate route to its new position, a place called Roffeno Musiolo near a prominent landmark outcropping called the Rocca di Roffeno.

In the first three days of the spring offensive, the 10th Mountain Division had nearly three hundred men killed in action and over a thousand wounded.

A busy road during the Push. Members of the press follow a jeep containing Mediterranean Theater deputy commander General Joseph McNarney and Fifth Army commander Lt. General Lucian Truscott. (*Signal Corps*).[c]

Road near Roffeno Musiolo. Note warnings of mines and booby traps. (*Denver Public Library*).[c]

The Rocca di Roffeno. (*Denver Public Library*).[c]

On April 17, Dad's battery fired a thousand rounds in support of the 87th Regiment's advance, which was picking up speed in a very fluid situation. Fifth Army commander General Truscott got intelligence the reserve German 90th Panzergrenadier Division, a mechanized-infantry outfit, was planning to attack the left flank of the 10th Mountain Division, at that time being covered by the Brazilians. The 10th, protected on the right by the 1st Armored Division, was taking the hard-fought high ground between the Reno River Valley and the parallel Samoggia Creek Valley about five miles west of the Reno. General Truscott decided to commit his reserve division, the 85th, to Highway 64 and the Reno River Valley, to free up the 1st Armored Division to both protect the 10th's left flank and use the more lightly defended Samoggia Creek Valley, rather than the Reno River Valley's Highway 64, to enter the Po.

The tanks of the 1st Armored Division were ordered to make the perilous crossing over the mountain roads, behind the 10th's advance—in essence moving west from the 10th Mountain Division's right to its left.

On April 18, the 1st Armored Division vehicles, traveling generally from east to west, had priority on the crowded mountain roads. The same day, Dad's battery moved north on Highways 25 and 26 through several towns, including the crossroads town of Tolè on Highway 26—from which the Apennines generally went downhill and the steep mountains gradually gave way to rolling country.

The noose was tightening on the Germans, not only in the Mediterranean Theater.

In the West, General Eisenhower's troops had crossed the Rhine and destroyed and captured the industrial Ruhr Valley, neutering the German war machine. Amid much controversy, he halted his armies at the Elbe River, only about sixty miles west of Berlin. The Western Allies were already past the area that was to be occupied by them when the war was over. Eisenhower viewed heavily bombed and mostly evacuated Berlin as only a political prize, not a military one, and not worth the estimated one hundred thousand casualties anticipated—especially in light of the continuing war in the Pacific. Over the objections of generals such as Montgomery and Patton, instead of going east to Berlin, he sent troops north toward the critical port of Hamburg and to Denmark—and south to Bavaria and Austria.

On April 16, while the Big Push was on in Italy, the Soviets on the Eastern Front began crossing Germany's Oder River with a million-man army, the first stage of the Battle of Berlin.

30

ANOTHER UGLY AMERICAN STORY

Exactly where and when this happened, I do not know; but I suspect it was during the perilous traffic movements while Fifth Army was in the process of breaking out of the mountains. I only know the incident, as Dad told it, made me very proud of him.

Dad's battery was passing through a devastated mountain town when suddenly the vehicles all came to a stop. An infantryman yelled, "Sniper!" and waved Dad to take cover. Dad grabbed his carbine and he and his driver got out of their kitchen truck. Dad explained sniping was a favorite tactic of the Germans to give them time to escape. They would leave a sniper behind in a church bell tower or other high position. When the sniper ran out of ammunition, he would surrender. Sure enough, the sniper soon dropped his empty weapon, waved a white flag and was brought to an infantry officer near Dad. The sniper apparently had killed and wounded several infantrymen. The angry, disgusted officer pointed to Dad who happened to be close by, holding his carbine. "Sergeant, kill the son of a bitch," he said.

Dad hesitated. "In cold blood, sir?" he asked. "I don't think I could sleep at night."

The first rule for an enlisted man is that when an officer gives a direct order, you obey. On the other hand, the rules of warfare were that a surrendering enemy was to be made prisoner, not killed. Dad said the officer gave him a dirty look, pulled out his pistol, and shot the surrendered sniper dead.

. . .

The night of April 18, Dad's battery set up near a place called Croce di Pradole, off of Highway 26, southwest of Montepastore, in the valley of Lavino Creek, a tributary of the Samoggia.

That night about 8 p.m., a rare thing happened. A desperate enemy plane strafed Dad's battery with machine-gun fire.

31

ARTILLERY

A word or two on artillery during the War to give you an idea of what the men in Dad's battalion and Dad's battery did. The 105mm artillery round actually consisted of several parts that could be changed by the men in the firing battery, depending on the mission. A special wrench enabled use of different types of fuzes, for example. The most common fuze was the super-quick percussion fuze, but some percussion fuzes had a delay feature, so detonation would occur a split-second after penetration of a roof or concrete structure. Other fuzes went off after a set amount of time. We already discussed the proximity or variable time fuze that was so devastating in 1945.

The projectile itself was usually filled with *high explosive* powder, called HE. Other projectiles were penetrating shape-charges, called *high-explosive anti-tank*, or HEAT rounds for use against enemy armor. White phosphorus chemical rounds were also used, as were illumination rounds to expose the enemy and smoke rounds to hide friendly activities. The fuzed projectile was placed in a cartridge. The cartridge had a built-in percussion primer and seven removable silk bags of various propellants strung together.

The fire direction center would tell the men at the firing batteries what fuzes to use and what type of round, as well as what numbered bags of propellant to use. Once the round had been prepared and loaded, and the barrel set at the proper deflection and elevation, the order to "Fire!" or "Fire for effect!" was given.

Moving a battalion, like Dad's, with its three self-propelled howitzer batteries, (18 M7's), thirty half-tracks, twenty-five 2 1/2-ton, twenty-one 1/4-ton and seven 3/4-ton trucks, most pulling trailers—not to mention the jeeps and specialized vehicles, like wreckers and three medium tanks for reconnaissance—was a formidable undertaking. The headquarters battery contained not only the command center, but also the battalion fire direction center, where the calculations were done—using maps, tables, and special slide rules—for giving firing orders to the howitzers.

In addition, headquarters battery contained survey teams and was responsible for the critical radio and wire communications essential to the operation of the battalion's missions, plus operation of the two observation aircraft.

The service battery was in charge of transportation, supplies (including food, fuel, and ammunition), as well as maintenance and repair of all the vehicles.

Bags of propellants for a 105mm artillery round.

Firing batteries, like Dad's, also had their own fire-direction centers, usually on half-tracks and could act independently, if required. The battalion fire-direction center was connected to the individual battery fire-direction centers by radio. Usually, the battery fire-direction center communicated with its six individual M7 crews by field telephone.

Accuracy of artillery fire, which operated in all weather, was dependent on numerous variables. The fire-direction center needed accurate contour maps showing elevations and needed to know the precise locations of the howitzers and the precise locations of the forward observers. In an age before GPS, this was not easy.

When a battery moved, the exact location of the battery had to be determined and the six howitzers put in parallel alignment. A delicate compass-like instrument on a tripod called an aiming circle was precisely leveled, magnetic north was determined with a sensitive built-in compass; and a known correction called the magnetic declination, specific to each location, was applied to find true north, as would be used on a map.

The normal horizontal 360 degrees of compass direction was not accurate enough. Instead, artillery used trigonometry to establish angles in a circle based on radians, rounded to a total of 6,400 milli-radians, or mils—with zero mils being true north. After an initial direction, or azimuth had been set, the deviation for a fire mission was called the "deflection." (The vertical setting, or "elevation," of the weapon was likewise measured in mils).

Mounted on each of the howitzers in the battery was a panoramic telescope instrument that had to be lined up precisely with a reflector on the aiming circle instrument and with two aiming stakes to the rear on the same line. Usually, all six howitzers in the battery were aligned in the same orientation; this was so instructions from the fire-direction center to the battery, if called for, could result in all six howitzers in the battery firing at the same target.

The "gunner," usually a corporal, on the left of the howitzer breech, controlled the deflection with the transverse hand-wheel and used the panoramic telescope to make sure the weapon maintained alignment after the recoil of every shot. (A recoil tube on top of the barrel used springs and heavy oil to minimize recoil movement). The assistant gunner, on the right side of the howitzer, controlled the elevation of the barrel with the elevation hand-wheel, opened and closed the breech for the ammunition loader, and pulled the lanyard to actually fire the weapon.

The versatile howitzer could fire at low angle for direct fire missions, if necessary, and at higher angles for its more common indirect fire missions, where the target was not visible to the cannoneers. The elevation of the howitzers above sea level was important and had to be taken into account, along with the elevation above sea level of the target. In some cases, for example, the howitzers might be in a valley and the target on a hill. The process of surveying for north and getting the battery's howitzer instruments calibrated for each move was called "laying in the guns." After that was done, the howitzers had to be "registered." Wind and weather had to be taken into account, a particular challenge in a mountain environment with its valleys and updrafts. One of the central howitzers in the battery would be directed to fire (usually a smoke round) at a known landmark observable by the forward observer. If the target was missed, the observer would tell the fire-direction center where the round landed, with reference to the target, so adjustments could be made until the landmark was hit. Those adjustments would then be figured into calculations for actual firing missions.

The amazing thing was that experienced batteries, like Dad's, in an emergency could actually set up and be ready to fire in a matter of minutes.

Dad told me the American infantry always tried to work under artillery protection. In the movies, the infantrymen usually attacked an enemy strong point, like a bunker or machine gun nest, with rifles and maybe hand grenades. In real life, Dad said, artillery usually would be called in to do the job.

One of the most devastating artillery tactics developed by the Allies, and perfected by the Americans, was called "time on target" or TOT. With its efficient communications system, an artillery battalion could mass the firepower of all its batteries to hit the same target at the same instant. Division artillery could even call for several artillery battalions to hit the same target at the same time (or within three seconds), giving the enemy no chance to dive for cover.

By 1945, with its integral observation aircraft, sophisticated communications systems, accurate fire direction, speed, mobility, flexibility, dedicated crews, and plentiful ammunition, some with the new variable time fuze, American artillery, arguably, had become the best in the world.

On the morning of April 19, due to the advances of the 10th Mountain Division, Dad's battalion was ordered to move forward to the vicinity of a place called Molina Oca. While an advance party was reconnoitering north of Montepastore for moving the battalion there, orders for a mission change came down from 10th Mountain Division Artillery.

The 85th Mountain Regiment had been ordered from the left side of the advancing 10th Mountain Division to the right. So, instead of supporting the 2nd Battalion of the 87th Regiment, Dad's battalion would be supporting the 1st Battalion of the 85th. The advance team from Dad's battalion had to adjust and move to the vicinity of a town called Amola. After a whole day of relocation, Dad's battery was finally in position and ready to fire at 8:30 that night.

32

FINALLY, FLAT LAND

April 20, 1945, was a momentous day. Dad's battery finally broke out of the mountains.

At 6 a.m., Company A of the 1st Battalion of the 10th Mountain Division's 85th Regiment, the unit Dad's battalion was supporting, pushed out to block a crossroads in the Po Valley. Private Thomas Hatfield, a scout for the 3rd platoon of Company A, became the first Fifth Army soldier to enter the Po Valley. The roadblock was established by 8:30 a.m. Within an hour, four of the six M7s in Dad's battery were rushed down Highway 26 to the roadblock on Highway 569, an east-west road in the very southern part of the Po Valley, to protect the infantry, which had come under enemy artillery and machine gun fire. The other two M7s were ordered to help lead troops of the 87th Regiment into a town called Crespellano, also on Highway 569, also out of the mountains, about five miles west of where the other four M7s had been dispatched. An hour later, the orders were changed and Battery B's M7s were to lead the troops into Crespellano with each Priest to carry six infantrymen, in addition to the regular crew. Dad's battery fired about six hundred rounds that day and reunited and set up at a place called Chiesa Nuova, just east of Crespellano, at dark.

Finally out of the mountains after so many months, Dad, for the first time, did something he long dreamed of—he slept in the Po Valley. But probably not for long, in the fast-moving combat situation.

The afternoon of April 20th, while Dad's battery had been busy supporting the infantry, scouts from the 10th Mountain Division's Company A of the 1st Battalion of the 86th Regiment had crossed the major east-west highway in the southern Po Valley, Highway 9, which ran from outside Milan over 150 miles to Bologna and was the major valley artery for the Germans to reposition and supply their troops in the Apennines. The advances made in the IV Corps zone with the 1st Armored Division on the left, the 10th Mountain Division spearheading in the middle, and the 85th Division on the right, exposed the enemy's flank in the II Corps zone to the east; and II Corps started to break through just south of Bologna. Fifth Army's General Truscott flew over the battlefield and made adjustments as to how his advancing troops would move on the flat ground of the Po Valley with its excellent system of highways. Truscott was not so much concerned with capturing cities, like Bologna, as he was with closing off enemy escape routes.

His secret plan to break through the German mountain defenses with the 10th Mountain Division west of the Reno Valley's Highway 64 had worked perfectly.

Meanwhile, the same day, in Berlin, ignoring the advancing Russian troops, Adolph Hitler, looking haggard, celebrated his fifty-sixth birthday in his underground Führerbunker with the top Nazi leaders. His left hand exhibited tremors of Parkinson's disease—or possibly tertiary syphilis. He also made his last public appearance by coming up to the ruined gardens of the nearby Reich Chancellery to pin medals on the boy-soldiers of the Hitler Youth.

To the surprise of many of his fellow Nazis, fearing capture if he tried to escape, Hitler announced his intention to stay in Berlin, no matter what. During a meeting with his military commanders and aware he could no longer conduct the war from a central location, he subsequently expanded the authority of Field Marshal Albert Kesselring, headquartered in Berchtesgaden, near Hitler's Berghof retreat in the Bavarian Alps, by placing Kesselring in charge of three army groups on the Eastern Front, two in Southern Germany, and, near the end, as we shall see, even Army Group C in Northern Italy.

Hitler still expected his soldiers to die for him and allowed his country and people to be destroyed. After Dunkirk in 1940, he had hoped Britain would surrender and become an ally of Germany in the fight against Communist Russia. Up to the end, he firmly believed it was only a matter of time before the Americans and British would go to war against the Soviets; and he was vainly holding out for whichever side would abandon the "unconditional surrender" requirement and offer him a deal to have Germany join the war against the other side. While the Nazi military commanders all knew a political solution was Germany's only hope, they also knew the Anglo/American-Soviet alliance, while intact, would never negotiate with Hitler. Several senior-level officers plotted to assassinate Hitler on July 20, 1944 at his Wolf's Lair Headquarters in East Prussia, but the briefcase bomb used resulted in only minor injuries to Hitler. He interpreted his miraculous survival as a sign he was to fulfill his destiny to destroy the Jewish-led Communists he claimed were a threat to all of Western Europe.

Hitler also had unrealistic hopes new weapons, like the single-engine jet Heinkel HE-162 Volksjäger or "People's Fighter," would turn the tide of the War. But the new weapons were insufficient in quantity to make any real difference.

While certain stresses arose among the Soviets, the British, and the Americans, as we shall see, the Allied coalition remained intact.

From the beginning of their alliance, even the British and Americans had somewhat different agendas. The fundamental policy of the Americans was to end the war as quickly as possible and bring the American servicemen home.

The British, on the other hand, had concerns about Communism, the makeup of post-war Europe, and holding on to far-flung territorial possessions. The Americans entered the War to save Britain from Germany, but they at times suspected they were being used to help Britain retain its colonial empire and influence.

As First Lord of the Admiralty in World War I, Churchill had been disgraced by a military debacle at Gallipoli in the Dardanelles Strait in Turkey. In 1943, when Churchill tried to get the Americans to attack Rhodes in a British campaign against the Axis-held Aegean islands near Turkey, leading again to the Dardanelles, he was told point-blank by General Marshall, "Not one American soldier is going to die on that goddam beach." (Churchill continued the campaign without American help, and Operation Accolade was a British disaster).

Like Hitler, Mussolini also had hopes the English and Americans would enlist the Germans and Italians to fight the seemingly unstoppable Red Army, which was conquering country after country in Eastern Europe. But, realistically, he knew his days were numbered. Mussolini had once met Winston Churchill many years before, and had exchanged secret communications with the anti-Communist British prime minister. As we shall see, these missives, or the lack of them, have spawned numerous conspiracy theories regarding the role of Britain in ending the war in Italy.

33

THE RACE TO THE PO

Intelligence sources supported the information previously given to the Allies by SS General Karl Wolff that once the Gothic Line was broken, the Germans would retreat north, make a stand at the Po River, and after that at the Adige River line, long under construction near Verona in the Alps, gateway to the Brenner Pass, and main escape route from Italy into Austria. Less certain were reports the Germans had a fortress in the Bavarian or Austrian Alps where they would make a last stand.

Fifth and Eighth Armies were not alone in their attack on the Po Valley. The partisans were busy making life unpleasant for the Germans in all the major northern cities: Milan, Turin, Genoa, Venice, Verona, and Bologna, among others. As the Germans abandoned their positions, the partisans moved in, usually ahead of the Allied armies.

The general plan of the Allies was to get to the Po quickly, trap as many of the enemy as possible south of the river, and then head for Verona and seal off the Brenner Pass, and other escape routes from Italy. All of the bridges across the Po River had long since been bombed by the Allies.

Allied armor in the streets of Bologna. Partisans had already taken over the city. April 21, 1945. (*Signal Corps*).[c]

The Germans would need to rely on boats to cross the river before the advancing Allied troops.

On the morning of April 21, Polish II Corps soldiers of the British-led Eighth Army, from Highway 9 east of the city, became the first Allied troops to enter Bologna, which had been abandoned by the Germans. (The Poles continued fighting valiantly in Italy, even though they felt betrayed when the Allies at Yalta decided Poland would be in the *Soviet* area of influence after the war). A few hours later, tanks and infantry of the 34th Division of Fifth Army's II Corps entered Bologna from Highway 65 to the south.

Bologna, for many months the objective of both Allied armies, had finally been taken.

After cutting off Highway 9, the next major obstacle for General Hays and the 10th Mountain Division, in the vanguard of Fifth Army's drive to the Po, was a tributary of the Po, the *Panaro River*, which stood between General Hays's troops and the most desirable ten to twenty miles for crossing the Po.

The Germans had prepared two bridges on the Panaro for demolition, one at a place called Bomporto and the other at a place a few miles east called Camposanto. The Germans delayed demolishing the bridges to allow as many as possible of their escaping troops to cross the Panaro River. The 10th Mountain Division, to which Dad's battalion was attached, had the mission of taking the Bomporto bridge, if possible. (The Camposanto Bridge was to be taken by the 85th Division, which was east of the 10th).

On April 21, while Dad's battalion was supporting the 10th Mountain Division's 85th Regiment at a place called San Martino in Casola, east of the Panaro River and south of Highway 9, 10th Mountain Division commander General Hays put together a task force to head toward the Bomporto bridge, about twenty-five miles from where Dad was. Task Force Duff, as it was called, was headed by Assistant Division Commander Brigadier General Robinson E. Duff and consisted of a battalion of infantry riding in trucks, the attached 91st (armored) Cavalry Reconnaissance Squadron, an engineer company, a light tank company, and a platoon of tracked tank-destroyers. By late afternoon, the task force succeeded in taking the bridge before the Germans could blow it. Orders were soon given to the rest of the division and supporting units, like Dad's, to head for Bomporto. Dad's battalion settled in near Bomporto at 11:30 p.m., about half an hour before the main body of the division arrived.

After taking the bridge intact, and crossing it, the task force had orders to continue on north to the Po River the next day.

Dad's battalion crossed the Panaro River at Bomporto, and on Sunday April 22 set up about ten miles west, near a town called Carpi, to protect the 10th Mountain Division troops. The main body of the division, and all three infantry regiments and support vehicles followed the task force and headed north toward the Po River and a town called San Benedetto Po, about twenty-four miles away.

Dad, I am sure, was unaware he would soon participate in one of the most bizarre incidents in the "fog of war."

Unfortunately, Task Force Duff, riding with no protection on its flanks, encountered mines, snipers, and delaying actions of one sort or another and was ambushed at San Benedetto Po by disorganized German troops armed with deadly Panzerfausts—shaped-charge, bazooka-like forerunners of the rocket-propelled grenade. The adversity was quickly

Bridge over Panaro River at Bomporto captured by Task Force Duff . April 21, 1945. (*Signal Corps*).[c]

overcome. Every crossroad presented danger, but the fighting was sporadic, and the task force, followed by the rest of the division, continued on.

The road to San Benedetto Po was strewn with disabled and abandoned German vehicles, as well as numerous dead German horses and dead enemy. Allied aircraft had destroyed many targets of opportunity as German units fled before the Americans. A lot of enemy vehicles were simply out of gas. An abandoned German motor pool was encountered. Since entering the Po Valley, 10th Mountain Division foot soldiers were driven forward by insufficient fleets of trucks that shuttled the soldiers for a few miles, dropped them off, and went back to bring more soldiers forward. Soon many of the infantrymen began traveling any way they could. Abandoned German trucks, staff cars, and even motorcycles were "hot-wired" by the weary marching men.

Numerous Germans surrendered—some of them gladly. They were disarmed and generally told to march, often unguarded, to the rear. Every ten miles or so, military police detachments escorted the prisoners back to "prisoner cages." The prisoners were fed American C-or K-rations (probably a disappointment after the food available to them in the Po Valley).

The advance into the Po Valley put a strain on Fifth Army's supply system. Dumps for fuel, food, and ammunition were quickly established around Modena (home of the Maserati automobile), west of Bologna. From the time the Big Push started for Dad on April 14, I suspect very little cooking was done by Dad's kitchen truck. The army on the move mostly relied on K-rations or C-rations. However, accounts I have read told of the reactions of the Italian people as the Americans came through their towns and villages. The war-weary people were obviously elated at the Germans being gone and offered the American troops who passed their way wine, grappa (brandy), and such food as fresh eggs. So possibly some fresh food was available for the troops who must have been sick of eating canned C-rations and pre-packaged K-rations.

As it progressed, the task force not only encountered scattered rear-guard delaying attacks, but also hastily-laid tank mines. When the task force was only about an hour away from the Po River, General Duff tried in vain to warn a tank driver, over the noise of the vehicle, of a mine. However, the tank hit the mine, crippling the vehicle; and shrapnel from the mine seriously wounded the general, who had to be evacuated. General Hays appointed his artillery chief, General Ruffner, to head the task force the rest of the way to the Po River, arriving about 6 p.m. on April 22, 1945. The remainder of the 10th Mountain Division was on the march during the night to assembly places near the south bank of the river.

Dad's battalion was also ordered to San Benedetto Po. In the darkness near a place called Moglia, Dad's battery (Battery A) and elements of the battalion's Headquarters Battery were involved in a strange nighttime incident.

On a straight stretch of open highway headed north, they passed a side road on the left. A smaller convoy, consisting of trucks and towing artillery pieces, turned from the side road, in the same direction as Dad's battery, sped up on the better road, and came abreast of Dad's convoy—the two convoys traveling side by side without lights. Some of the guys in Dad's convoy yelled a few catcalls at the other guys, who were uncharacteristically quiet. Even in the darkness, driving without headlights, the guys in Dad's battery and Headquarters Battery sensed the outlines of the vehicles beside them were unfamiliar.

It was soon apparent the other convoy was German!

Both convoys stopped. The Americans grabbed their weapons, jumped out of their vehicles, and quickly captured the entire enemy convoy, including two large 170mm artillery pieces, vehicles, equipment, and about a hundred German soldiers, who were sent to the rear as prisoners. The Germans had intended to cross the Po at night, if a boat could be found, since all bridges were out, and Allied aircraft flew over the area in the daytime. Dad's battery had only minor casualties in the incident.

Dad and his battery continued on, setting up in the darkness of the early morning hours of April 23, just south of San Benedetto Po, near Headquarters Battery, with Batteries B and C north of the town, closer to the south bank of the Po River.

34

CROSSING THE PO

As the 10th Mountain Division approached the river, General Hays knew crossing the Po, where all bridges had long been taken out by Allied air power, would be his next objective. An amusing story about this has become lore in the division.

The truth is that the breakout into the Po Valley was expected to be in the II Corps zone and not in the IV Corps zone, so all of Fifth Army's boats and bridge equipment had been assigned to II Corps. When the 10th Mountain Division broke through the Gothic Line and became first to enter the Po Valley, Fifth Army commander General Truscott had to reassign the river-crossing equipment, which was far to the rear and the east—much of it on a train south of Bologna—to IV Corps; and getting it to the Po River would take some time. General Hays and his men got to the river sooner than anyone expected, and General Hays wanted to cross the river ahead of anyone else, certainly ahead of as many retreating Germans as possible, and before enemy reserves could be called in to defend on the other side.

So, as General Hays approached the Po River on April 22, he ordered his Engineer battalion (the 126th Mountain) to send a detachment to the engineering supply depot in the rear to find him some *boats*. A warrant officer named Heller was put in charge of getting the boats. Sixty miles east, enroute to the supply depot, Heller and his detachment came upon a convoy of five trucks, each hauling ten 16-man M-2 wooden assault boats, a total of fifty boats. Heller was told the boats were intended for "the 85th," meaning the 85th *Division*. Unlike the 10th Mountain Division, which had quickly captured the Bomporto Bridge, the 85th Division, to the east of the 10th, had encountered heavy contact at the other bridge across the Panaro River at Camposanto and had fallen behind. "Oh, the 85th," said Heller, "they need the boats badly, just follow us." Heller later said he, of course, was talking about the 85th *Regiment* of the 10th Mountain Division, not the 85th *Division*. The boats with paddles arrived at the Po River about 8 a.m. on April 23rd and were quickly unloaded, after which the convoy commander was sent on his way—not knowing how he had been had.

In the morning on April 23, Dad's battery and the Headquarters Battery moved east of San Benedetto Po, closer to the other batteries and in position to protect the planned river crossing by 10th Mountain Division troops. Additional firepower was provided by the BL 5.5-inch (140mm) guns of the British 178th (Lowland) Medium Regiment of Royal Artillery, which, like Dad's battalion, had been attached to the 10th Mountain Division to supplement the division's smaller and lighter pack artillery.

H-hour was set for noon. Risk-taking General Hays, hoping to get his troops across the river before the Germans could mass a defense on the opposite shore, sent the 1st Battalion of the 87th Mountain Infantry as the lead element. The assault boats each carried about a dozen soldiers with gear, plus three engineers from the division's 126th Mountain Engineering Battalion—having no training for such an operation—who paddled, two in back and one in front. The crossing immediately came under enemy fire.

The sandy banks of the river were a little less than a thousand feet apart, about the length of three football fields. The Americans were lucky. The Po usually was much higher in April due to spring rains, but actually the river was at its lowest point for April in half a century. This made the river not only narrower and shallower than expected, but also

87th Mountain Infantry Regiment of 10th Mountain Division crossing the Po River in assault boats on April 23, 1945. (*Denver Public Library*).

the current was more moderate, an important consideration for paddling boats from one side to the other. The Germans had not counted on the river being so tame. Deep-draft boats they intended to use to get their armor and heavy equipment across in retreat were useless in the shallower river.

The German efforts to cross the river before the Americans, and set up a line of defense, had a number of other problems. German units in retreat from the mountains often left their engineers behind to lay mines or otherwise devise means of delaying the American pursuit. However, at the river, those engineers were needed to help in crossing. German communications were disrupted as the chain of command broke down. Units cut off from their headquarters frequently had to fend for themselves. Dug in, up in the mountains, the Germans could hide from Allied air power. In the flat land of the Po Valley, hiding was much more difficult.

Potential crossing places along the Po had been studied by Fifth Army for months based on aerial photographs and other intelligence. The crossing location north and east of San Benedetto Po was wisely chosen. At that point, a tributary of the Po, the Mincio River, was only a mile or so north of the Po. A German force defending against the crossing was unlikely since it would have been caught between the two rivers.

I assume aerial reconnaissance convinced General Hays that while the surprise crossing by his troops starting April 23rd would result in casualties, the casualties would be minimal compared to if the Germans made a stand at the river. He knew that with his troops across the Po, a German defense line on the north bank of the river would be impossible to maintain and likely would not even be established, saving many Allied lives.

During the entire time of the crossing, from noon on April 23rd, through the night and into the following morning, there was sporadic enemy fire from artillery, mortars, and machine guns. Particularly frightening were air bursts from 88mm Flak anti-aircraft guns firing horizontally with time fuzes to try to stop the repeated trips by the assault boats. Battery B of Dad's battalion, which had been close to the shore, was heavily shelled by enemy artillery, wounding an officer and eight enlisted men. The Allied howitzers fired everywhere the enemy was suspected. Dad's battery alone fired a thousand rounds to protect the crossing. As the little boats made more and more crossings, more and more of the 87th was able to secure a bigger beachhead. By nightfall, the entire 87th and 85th Regiments had crossed. Medical detachments had been taken across to care for the wounded on the north shore, as well as two battalions of the light 75mm pack howitzers. The 86th Mountain Infantry Regiment crossed during the night.

General Truscott, Fifth Army commander, was busy on April 23. After spending the previous afternoon in a victory parade in Bologna with Fifteenth Army Group commander General Clark, IV Corps commander General Crittenberger, and II Corps commander General Keyes, as well as various Eighth Army commanders, he was eager to get back to leading his troops.

The morning of April 23, General Truscott flew from his headquarters to an airfield near Florence in an L-4 Piper Cub observation plane piloted by General Thomas C. Darcy, commander of Twelfth Air Force's XXII Tactical Air Command, which provided tactical air support for Fifth Army. There, he and General Darcy transferred to a two-seat variant of a P-51 Mustang fighter plane to do reconnaissance over the battle area.

The British-designed, North American Aviation-built, P-51 with its revolutionary wing design and Merlin engine, licensed in the US to Packard, able to operate at high altitudes and with a top speed of 440 mph—fastest in the Allied arsenal—shot down nearly six thousand enemy aircraft during the War, the most of any Allied fighter.

However, it was 100 mph slower than the twin-engine German ME-262, the world's first functional jet fighter, which became operational too late and in too few numbers to change the outcome of the War. ME-262's were not known to have been used in Italy, nor were British Gloster Meteors, the only Allied jets. (If Dad saw a jet in northern Italy near the end of the War, it would have been the German AR-234, the world's first operational light jet bomber—again too few and too late—used in Italy by the Germans for high altitude reconnaissance).

To Generals Truscott and Darcy, the German forces they observed on the ground in the southern Po Valley appeared to be in a state of confusion and disorder, attempting to avoid the oncoming Allied troops. General Darcy strafed a column of German vehicles west of San Benedetto Po. Several other German columns were seen approaching the river somewhat east of San Benedetto Po, at a place called Ostiglia, which had been chosen as another crossing location by the Allies, but where Allied troops had yet to arrive. General Darcy did some more strafing and called for tactical air support to attack the German convoys in the area.

General Darcy flew General Truscott back to Florence, and they transferred again to the Cub observation plane to fly to General Hays's 10th Mountain Division headquarters. They arrived in the afternoon, while the 10th Mountain Division was in the process of crossing the Po in the assault boats. Landing in a wheat field, General Darcy actually cracked up the small plane, but neither general was hurt. When General Truscott arrived at General Hays's headquarters trailer just south of the river and learned most of two regiments had already crossed the Po, he supposedly asked General Hays, "What the hell are you trying to do, George, win this war all by yourself?"

General Truscott was now more aware than ever of the urgency of getting bridging equipment to IV Corps and to San Benedetto Po without delay—so tanks, supply trucks, and other vehicles could cross the river as soon as possible and continue chasing the Germans.

Something that indirectly was to affect Dad also occurred at the meeting. General Hays was being visited that day by one of his former pre-War battery commanders, a friend who was one of the most famous

DUKW amphibious vehicles (called "ducks") arrived near San Benedetto Po on April 24, 1945 to transport supplies across the Po River for 10th Mountain Division troops who had crossed the day before in assault boats. (Some of the loading and unloading was done by German POWs (in suspenders). Note small assault boat offshore in the middle of the picture.) (*Denver Public Library*).[c]

soldiers of World War II, Colonel William Orlando Darby. Colonel Darby happened to be in Italy as part of a touring group of brass, including the head of the Air Force, Five-Star General Hap Arnold.

General Hays told General Truscott he had lost his assistant division commander, General Duff, who had headed Task Force Duff.

He asked if Col. Darby could be appointed his new assistant division commander to replace the wounded General Duff. General Truscott said he would make the official request of 15th Army Group commander General Mark Clark and General Joseph McNarney, top American in the Mediterranean Theater. In the meantime, General Truscott, who also knew Col. Darby well, told the colonel to consider himself appointed.

More on Col. Darby later.

By the time a convoy from II Corps containing heavy bridge matériel, sixty amphibious DUKW boats, rafts, storm boats, and fifty more assault boats arrived at the crossing area the morning of the next day, April 24th, all three infantry regiments of the 10th Mountain Division had crossed and established a bridgehead perimeter on the opposite shore.

Only twelve of the original fifty assault boats were left. Some engineer crews had made more than twenty round trips in the heavily-laden,

This M-1 pontoon steel treadway bridge, completed the morning of April 25, 1945, was the first bridge Fifth Army built across the Po. It could carry trucks and light vehicles. (The officer in the soft hat is IV Corps commander Lt. General Willis Crittenberger. The two-star general to his right is Maj. General George Hays, commanding general of the 10th Mountain Division. To the right of General Hays is probably Maj. General Vernon E. Prichard, commanding general of the 1st Armored Division. (*National Archives*).[c]

easily-swamped boats under fire. Two engineers had been killed and twenty-two wounded, but a whole division had crossed the river, making possible the building of bridges.

It was another amazing achievement by the 10th Mountain Division.

Dad's battalion fired two-hundred rounds on April 24th, but enemy activity in the area had been mostly neutralized by the artillery and tactical air presence.

By afternoon, construction had begun on a pontoon steel treadway bridge. Less than three miles upstream, a steel and timber Bailey pontoon bridge was being erected, capable of carrying tanks and heavy trucks. Work on the two bridges continued through the night.

By about 10:30 in the morning on April 25, the Treadway bridge, spanning about 915 ft., was completed. Only a few hours later, the Bailey bridge, about 840 ft. long, was ready.

The Germans had tried to blow a bridge at nearby Governolo on the much smaller Mincio River, a tributary just north of the Po, but most of it seemed intact. Several hours were lost as Allied engineers evaluated the bridge. A tank, crossing at slow speed, demonstrated the bridge could handle heavy traffic, and Allied armor and other vehicles were soon crossing both the Po and the Mincio Rivers.

The Germans had anticipated the Allies would outrun their supply lines and stop at the Po. Instead, Fifth and Eighth Armies had formulated an intricate transportation system to keep the advancing Allied troops well supplied with fuel, ammunition, and food all the way to the

Combat engineers working on the other bridge across the Po, a British-designed heavy Bailey pontoon bridge, less than three miles upstream (i.e. west) of the treadway bridge, also finished on April 25, 1945, only a few hours after the treadway bridge. This bridge could carry armored vehicles, like the M7 Priests, and is probably the one Dad's battery crossed the next morning. (*National Archives*).[c]

Po—and beyond. The Germans were to be given no opportunity to regroup on the other side of the river.

About 6 p.m. on April 25, Dad's battalion received orders to move to an assembly area along the river for the night, west of a place called Camatta, and prepare for crossing the river first thing the next morning.

35

THE UPRISING

While Dad and his battalion with their heavy M7 Priests were awaiting a scheduled time for crossing the Po, the Allied advances to the river had already resulted in much activity in Northern Italy, of a political, as well as military nature. When spearhead Polish troops from Eighth Army entered the critical Po Valley transportation hub of Bologna from the east on April 21, followed a few hours later from the south by troops from Fifth Army's 34th Division, much of the city was already in the hands of the *partisans*, political militias.

Outlawed anti-Fascist and anti-Nazi political groups, such as the Italian Communist Party, Italian Socialist Party, the Action Party (republican socialist), the Christian Democrats, Liberals, and others, had organized against their common enemy as the Committee of National Liberation of Northern Italy (*Comitato di Liberazione Nazionale di Alta Italia,* in Italian) called the CLNAI.

Militia "brigades" of the CLNAI's military arm, the CVL (Corpo Volontari della Libertà) fought the Nazis and the Fascists of the Italian Social Republic, the puppet government set up by Mussolini after his rescue by the Germans. The Italian Social Republic was really run by the German ambassador, Rudolf Rahn, and the SS general in charge of Italy and Austria, Karl Wolff.

The CLNAI collaborated with the Allies and with the king's government, now under new Prime Minister Ivanoe Bonomi. The Americans and the English air-dropped arms, supplies, radios, even money—as well as human operatives—behind the lines for the partisans and gave the CLNAI resistance 160 million lire per month.

Of utmost concern were the three major western cities of the so-called Italian Social Republic: Genoa, the largest seaport in Italy; Turin, the home of Fiat and major industrial center; and Milan, Italy's largest metropolitan area—all of which had been bombed numerous times by the Allies during the course of the war. The CLNAI called for a general strike on April 24–25, 1945, shutting down all the factories and transportation facilities, and warned the Germans and Fascists to "Surrender or Perish."

The afternoon of April 25, Benito Mussolini, an avowed atheist, came to the palace of the revered Archbishop of Milan, Cardinal Ildefonso Schuster, who had offered to negotiate with the Allies on his behalf. The Cardinal wanted to prevent the destruction of Milan and the Lombardy region, where Mussolini had previously threatened to make a last stand. He set up a meeting between Mussolini and the military arm of the CLNAI partisans. Mussolini wanted safe passage the following day for himself and three thousand of his paramilitary Black Brigade soldiers from Milan to a place called Valtellina, a valley in the Alps north and east of Lake Como. Valtellina had been fortified with pillboxes, gun emplacements, and man-made caverns during World War I, when Italy was at war with Austria. Advised by Alessandro Pavolini of the Black Brigades, Mussolini said he now intended to make his last stand there instead of Milan.

With the strike on, Germans beginning to evacuate the city, and with newspapers and radio stations falling under partisan control, the partisan insurrection was already under way—not only in Milan, but also in Turin and Genoa and other cities in the Po Valley.

In front of the Cardinal, the CLNAI representatives told Mussolini, the partisans, like the Allies, would accept only unconditional surrender. All parties who surrendered would be prisoners of war and would

Much-respected Cardinal Ildefonso Schuster, *beatified* by the Roman Catholic Church, inspecting Allied bomb damage to the roof of the Milan cathedral (or Duomo) earlier in the War.

be treated as such. Officially, the partisans were unwilling to guarantee Mussolini and his most loyal followers safe passage through partisan-controlled areas.

After Mussolini left, a German delegation visited with the Cardinal and confirmed that while SS head Wolff favored the unconditional surrender terms of the Allies, word had yet to come from General Heinrich von Vietinghoff on the army's position. The Germans wanted safe passage north through partisan-held areas, offering to forego the demolition and damage they planned to do to Italian cities, to buy time for their escape from the Fifth Army.

The partisans could not officially sanction letting the Germans escape. However, it was in their interest to avoid unnecessary confrontation, particularly if their infrastructure would be left intact.

Some accounts say Mussolini left Milan without responding to the partisan demands for surrender; others say he left word of his refusal.

After the meetings at the Cardinal's palace—first with Mussolini and then with the Germans—the CLNAI representatives told the Cardinal that with the Fascists and Nazis now in flight, they, the partisans, were taking over the government.

Partisan uprising in Milan's Piazza del Duomo. April 25, 1945[c]

Thus, though the war was not over, **April 25** became the day commemorated in Italy as Liberation Day—even now a public holiday.

That evening, the new "government" issued a decree calling for execution of the top Fascists, including Mussolini. Partisan roadblocks were set up to keep Fascists—though not necessarily Nazis—from fleeing the country. Unlike the Allies, who were trying to keep as many Germans as possible—and their weapons and equipment—in Italy, the partisans unofficially wanted as many Germans as possible to leave. The CLNAI knew, however, their administration was temporary. They had previously agreed with the Allies to hand over power to the Allied Military Government whenever it arrived in Milan.

. . .

Elsewhere on April 25, over three hundred Lancaster heavy bombers from Britain's Royal Air Force, flying from a future post for Dad, Foggia in Southern Italy, conducted a major bombing raid on Berchtesgaden in the Obersalzberg of the Bavarian Alps, destroying the *Berghof*, Hitler's alpine chalet, as well as nearby chalets of Hermann Göring, Martin Bormann, and other top Nazis.

The threatened German last stand at some unknown fortress in the Alps was beginning to look more and more like what it really was—a fiction.

36

THE CROSSING

At 5:30 the morning of April 26, Dad's battalion crossed the Po River, arriving at an assembly area west of a place called Casale at around 7 a.m. Casale was across not only the Po but also the Mincio River and a canal. I assume Dad and his battalion crossed the Po on the heavy pontoon Bailey bridge since that bridge seems to have been closer to where they were before and after crossing, and was designed to take armored vehicles, like the M7 Priests. Within twenty-four hours some 3,400 vehicles are believed to have crossed the river on the Bailey and the M-1 treadway bridges, including not only support vehicles for the 10th Mountain Division, but also the 1st Armored Division and part of the 85th Division.

With the two bridges completed, all three advancing divisions of IV Corps—the 1st Armored Division on the left, 10th Mountain Division in the middle, and 85th Division on the right—were operating north of the Po. The Brazilians, meanwhile, and the 34th Division, recently assigned from II Corps to IV Corps, were busy mopping up the thousands of German troops still south of the Po. On the coast, the 92nd Division had taken La Spezia, finally silencing the coastal guns at Punta Bianca that had so long hindered any advance toward Genoa. Everywhere, the German resistance was collapsing.

The same day Dad's battalion crossed the Po, a third bridge, a heavy M-2 treadway, was completed over the Po River by Fifth Army about fourteen miles east of San Benedetto Po, at Quingentole, near Ostiglia. Similarly, the British-led Eighth Army, east of Fifth Army, had constructed bridges and was advancing north of the Po. The unanticipated speed with which the Fifth and Eighth Armies bridged the Po left numerous enemy units trapped south of the river.

North of the river, the race was on to prevent the Germans from making another defensive line, in the Alps, and to keep them from retreating into Austria.

37

A PICTURE I DO NOT HAVE

I don't remember all of the pictures of his war experience that Dad showed me as a boy. One does stand out, however. I remember it because of its content: a lot of nude men bathing in a river. I have a vague recollection of Dad saying the men were so happy to reach the Po River, which had been an objective since the beginning of the campaign.

The US Army emphasized hygiene. All men were to be clean-shaven. Helmets became water basins, and the GIs used safety razors and brushless shave cream to shave *every day,* if possible. (Dad was more partial to the Schick-type injector single-edge razor than to the double-edge Gillette-type). Of course, daily shaving was often impossible when in actual combat.

Accounts I have read of the Big Push out of the mountains suggest the American troops were almost constantly on the move, sleeping when and where they could, eating only C-and K-rations, with almost no time to brush their teeth, shave, bathe, change or wash out their socks, underwear or fatigues, or otherwise take care of their personal needs.

So, it is understandable the men would take a bath in the river if they could.

I do not think Dad was in the picture, and I do not recall the scene as being in a broad stream like the Po. So, maybe it was taken at the Mincio River, or maybe at some other place and time.

Anyway, I have not been able to find the picture. Maybe Dad was concerned about the full-frontal nudity and threw it out somewhere along the line. Or maybe it just got lost in a move sometime over the years. Anyway, the lack of that picture makes me admit I do not have all the photos Dad brought home from his War years.

I wish I did.

38

TASK FORCE DARBY

With his new assistant division commander appointed, General Hays named him to head another, larger task force to lead the 10th Mountain Division toward its next objective—Verona in the foothills of the Alps.

Dad's battalion was chosen to be part of the new *Task Force Darby.*

Colonel William O. Darby had become a legend as leader of what was known as *Darby's Rangers.* A 1933 West Point graduate, Darby had served as an artillery officer under General Hays back in the States, but also had commanded a cavalry unit and had amphibious training in the Caribbean. In 1942, Major Darby found himself in England when General Truscott, then charged with forming an American special operations

Colonel William O. Darby.[c]

team, appointed him head of the new unit, which was to be based on the British *commando* model. General Truscott wanted the American version to be distinguished from the British, so he called members of the new battalion-sized unit "Rangers." Darby had his pick of the only two divisions in England at the time, the 34th Infantry Division and the 1st Armored Division, and trained his Rangers, numbering about 450 men, in Northern Ireland and Scotland, alongside the British commandos. Success of the Ranger battalion in North Africa led to additional battalions under Darby. The Rangers distinguished themselves in the Sicily invasion. Unfortunately, Darby's Rangers were decimated at Anzio in 1944, and Darby was sent to Washington for war planning. Darby was awarded numerous medals including the Distinguished Service Cross, Silver Star, and Purple Heart. He was noted as a great organizer who also led his men in battle. Although his organizational skills were valued in Washington, he was eager to get back into action since Anzio. After the War, a Hollywood movie was made about him called *Darby's Rangers* starring James Garner.

Besides Dad's battalion (the 1125th Armored Field Artillery Battalion), Task Force Darby consisted of the 10th Mountain division's 86th Mountain Infantry Regiment, the 13th Tank Battalion (attached from the 1st Armored Division), two companies of tank destroyers, two battalions of the division's pack artillery, and elements of the division's engineer battalion with bridging equipment.

Actually, three divisions from Fifth Army were now headed north toward Verona, the 10th Mountain Division and 85th Division from IV Corps and the 88th Division from II Corps. Aside from the tank battalion that was part of Task Force Darby, the 1st Armored Division arced west to prevent the enemy from entering the Alps and was well on its way to Lake Como and Milan.

While the 10th Mountain Division, traveling Highway 62, succeeded in capturing the Villafranca Airport about ten miles outside of Verona and waited for the armor and self-propelled artillery to catch up, parts of the 85th Division and, the 88th Division, crossing at the bridge over the Po at Ostiglia, went straight to Verona on Highway 12. They arrived at about 9 p.m. on April 25 to find much, but not all, of Verona, city of *Romeo and Juliet*, in the Adige River Valley, leading to the Brenner

Pass—main German escape route to Austria—already in the hands of the partisans.

By the time Dad's battalion crossed the Po in the early morning hours of April 26, it learned most of Task Force Darby had already left the crossing area the night before. By 8 a.m. Dad's battalion was on its way to catch up. At around noon, the battalion had reached a place called Dossobuono, about seven miles outside of Verona and not far from the Villafranca Airport, where American aircraft were to begin operating the next day. The battalion halted for about two hours awaiting further orders. While some elements of Task Force Darby had made it inside the walled part of Verona by about noon, it was decided the 85th Division would occupy the city and prevent the Germans from manning their so-called Adige Line on the Adige River. The 88th Division was to head for Vicenza, east of Verona.

As for the 10th Mountain Division, General Hays decided the shortest route to the Brenner Pass, where the Germans hoped to escape into Austria, would be several miles west of Verona, along the east shore of Lake Garda, which was surrounded by mountains. So, Task Force Darby and most of the 10th Mountain Division headed toward the lake. In the rain, Dad's battalion went into position southeast of Bussolengo, located between Verona and Lake Garda. Dad's battery fired 150 rounds that day "with good results."

If and when it reached the north end of Lake Garda, the 10th Mountain Division would have a choice of taking a smaller highway directly north toward Trent or going east about five miles to the Adige River Valley with its main highway and rail line, also leading to Trent, about thirty miles north of the lake. About forty miles north of Trent, on the main highway to the Brenner Pass and Austria, was the major objective of Bolzano. The Alpine city of Bolzano—Bozen, in German—was in South Tyrol, an Italian region that had been part of Austria before the First World War. The majority of the inhabitants still spoke German as a first language. Bolzano was where the German Army had its new headquarters in Italy.

Located at the foot of the Alps, Lake Garda—over thirty miles long, ten miles wide near the southern end, and less than two miles wide at its northern end—was Italy's largest lake. In 1945, it had political significance.

Mussolini's puppet government, the Italian Social Republic, had its chief offices in Salò on the southwest shore of the lake, near Gargnano, where Mussolini's home, Villa Feltrinelli, was located. Many people called the government the Republic of Salò. Mussolini's mistress, Clara Petacci, lived in a nearby villa on the west shore of the lake, as well.

In the drive up the east shore of Lake Garda, Dad would be on the move a lot and would not get much sleep. At 6:15 in the morning on April 27, 1945, Dad's battery moved from Bussolongo outside of Verona about four miles northwest to a place called Pastrengo, arriving there and setting up in about an hour and a half. By noon, Dad's battery had moved again and was right near the eastern shore of the lake between Lazise and Bardolino. In the afternoon the howitzers from Dad's battery provided "destructive and neutralizing fire" on holdout enemy tanks and guns in the mountains east of the lakeshore. By 8 p.m. the battalion had moved north on the lakeshore highway and established positions in a town called Garda. Shortly after midnight, they were on the move again. At about 4 a.m. on April 28, Battery A, Dad's battery, set up in the town square in Malcesine, about ten miles south of the north end of the lake, and fired on enemy positions and personnel, often in direct fire missions, where enemy targets could be seen. At one point, according to Mr. Warner's book, the six M7s in Dad's battery were pointed in three different directions at the same time.

To keep his 10th Mountain Division troops fresh, General Hays used a program of rotating or leapfrogging his three infantry regiments, the 85th, 86th and 87th, into Task Force Darby—eight hours of fighting followed by sixteen hours in the rear. So there was a lot of activity on the road along the east shore of Lake Garda.

After April 28, the 10th Mountain Division, and Dad, whose battalion was attached to it, were removed from Fifth Army's IV Corps, now operating mostly in the western Po Valley—under whose command they had been since arriving in Italy—and placed directly under Fifth Army control.

On April 29, early in the afternoon, Dad's battery moved farther north along the shore, just past a town called Navene, about five miles from the north end of the lake. At a distance north from that point, the leading elements of Task Force Darby had come to a stop the previous day. German demolition had blocked the first of six or more tunnels on the lakeshore highway where the 7,000-foot mountains came right down to the lake. Dad's battery was set up to protect the troops as DUKW amphibious vehicles operated out of a cove near Navene to ferry the advancing infantry, under fire, around the sabotaged tunnel.

Self-propelled German 88mm Flak anti-aircraft guns in the mountains across the lake lowered their barrels and used mechanical time fuzes to fire air bursts on the soldiers in the DUKW vehicles. Dad's battalion did its best to neutralize the enemy fire. The fighting seemed to be getting tougher while chasing the retreating Germans, as the 10th Mountain Division was in the mountains once again.

The blocked tunnel and damage to the road and a bridge beyond it stopped any further advance by the M7 Priests in Dad's battalion until the engineers could get to work with their bulldozers.

But that did not stop Task Force Darby.

Having used the amphibious DUKWs past the blocked tunnel, the infantry continued its pursuit, taking each tunnel along the lakeshore road as it came. The immediate objective was the town of Torbole, past the tunnels and at the northeast end of the lake.

U.S Army Signal Corps photo of Battery A (Dad's battery) firing direct fire mission across northern Lake Garda on April 29, 1945.[c]

10th Mountain Division troops in a Lake Garda tunnel with pack artillery that had been transported by DUKWs. (*Denver Public Library*).[c]

Suddenly, a giant explosion was heard ahead.

At the entrance to the second-last tunnel, about 500 feet long, the foot soldiers encountered machine-gun fire coming from a slit in the rock face, which they overcame with a grenade, killing eight Germans.

When they entered the tunnel, the Americans realized the explosion they had heard was a German demolition attempt gone wrong. An estimated forty German soldiers were dead inside—from an explosion meant for the 10th Mountain troops. After going through the hellish carnage and gore in the south end of the tunnel, the Americans were halted at the other end by enemy machine gun, mortar, and artillery fire. A number of the soldiers had sought refuge in the tunnel when a lucky hit from a German 88 gun, an air burst inside the tunnel, killed seven and wounded about forty of the 10th Mountain Division troops.

Nevertheless, relying on tactical air support, the longer-range guns of the 178th (Lowland) Medium Regiment of Royal Artillery, and the portable pack artillery brought by DUKWs, plus supporting fire from Dad's battalion, Task Force Darby relentlessly pressed on toward Torbole in the face of a determined enemy.

Meanwhile, events elsewhere were beginning to overtake the fighting at Lake Garda.

39

THE CAPTURE OF MUSSOLINI

Accounts of the final days of Benito Mussolini are often in conflict—one source says there are at least nineteen different versions of how he was arrested and died—though the gruesome public spectacle at the end was well documented. Conspiracy theories abound. The story has been told different ways in numerous books and magazine and newspaper articles—many not available in English—and in several movies. (Mussolini has been played on film by actors such as Rod Steiger, George C. Scott, and Bob Hoskins).

Pardon me if I go on a bit at length about Mussolini and his dramatic end, but the story, at least as I see it, is one of the most interesting in modern history—and took decades to be revealed.

And it ties into a photo Dad brought home.

. . .

Italian Fascism espoused a doctrine called *spazio vitale*, or "living space," which meant Italy supposedly needed to spread its population and culture—an excuse for empire building. This was the underlying basis for

Italy's conquest of Ethiopia in East Africa and colonization of Libya in North Africa in the 1930s. The grand vision was for Italy one day to call the Mediterranean *Mare Nostrum*, or "our sea," as the ancient Romans had in classical times. Similarly, Hitler's political philosophy for the Nazis revolved around *Lebensraum*, or "room to live," claiming the German people needed to spread out. The Fascist philosophy, however, was not so racist as the German. Although the Italian Fascists committed atrocities, like the use of poison gas in Ethiopia, and was unquestionably anti-Semitic, actual genocide was not part of Fascist doctrine, as it was for the Nazis, and was generally abhorrent to the Italian character. Italians were proud of their "mongrel" (European and Mediterranean) heritage and mocked the German belief in a *pure* Aryan race as a fanciful myth.

During the 1930s, Italy and Germany, enemies during World War I, gradually became allies. However, Mussolini was upset with Hitler when Austrian Nazis, under orders from Hitler, assassinated the diminutive Fascist Austrian dictator Engelbert Dollfuss in a coup attempt in 1934 after Mussolini, who became personal friends with Dollfuss, had hoped to make Austria a buffer between Fascist Italy and Nazi Germany. Mussolini, at the time in his Adriatic villa hosting the wife and children of Dolfuss, personally informed the widow of her husband's death. Taking the assassination personally, he threatened war with Germany. Hitler, surprised by Mussolini's reaction, denied all responsibility.

In order to support his ambitions in the Mediterranean, and East Africa, Mussolini built a considerable navy, which is really what brought Italy into conflict with its former World War I allies, France and Britain. In 1935, to deter Italy's incursion into Ethiopia, the League of Nations, led by France and Britain, imposed a crude oil and coal embargo on Italy. This made Italy almost totally dependent on Germany for coal (and on Mexico for oil) and hit at a time when Mussolini's costly forays into Africa had devastated the Italian economy. When Germany became the only major European nation not to condemn Italy's conquest of Ethiopia, Mussolini gave up on protecting Austria from the Nazis. In the late 1930s, Germany and Italy cooperated in assisting the "Nationalist" cause of Francisco Franco in Spain's civil war, where both Axis countries experimented with new weapons and air power tactics—though Italy had little to spend on modernizing its own armed forces.

In May of 1938, Germany relatively peacefully annexed *German-speaking* Austria, Italy's enemy in World War I and neighbor to the north. Hitler next set his sights on the Sudetenland, a region of Slavic western Czechoslovakia *with many German speakers*, part of the former Austro-Hungarian Empire. (Mussolini secretly feared the Nazis might next attempt to take back Alpine Italy's *German-speaking* South Tyrol region and the Adriatic port of Trieste—former Austro-Hungarian possessions awarded to Italy as a consequence of World War I).

In late September of 1938, Italy, Germany, France, and Britain signed the "Munich Pact" on the promise by Germany of no further expansion in Europe beyond the Sudetenland. Conservative British Prime Minister Neville Chamberlain excitedly told the press he had avoided war and negotiated "Peace for our time." Chamberlain's "triumph," however, was criticized by Winston Churchill, who famously said, "You were given the choice between war and dishonor. You chose dishonor, and you shall have war." Churchill was right. Six months later, in March of 1939, Hitler violated the pact by invading the rest of Czechoslovakia.

In April of 1939, not to be outdone by its ally, Italy crossed the Adriatic Sea and conquered Albania in five days, giving it a foothold for future aggression against Yugoslavia, to the north, and Greece, to the south.

In May of 1939, Hitler and Mussolini signed a military cooperation treaty, popularly called the *Pact of Steel*. Japan, like Italy, had been on the side of the Allies in World War I, and was supposed to sign, too, but Hitler and Mussolini were concerned with Britain and France, whereas Japan wanted the treaty to address the Russians; and Hitler was secretly negotiating a nonaggression pact with the Russians over his next objective, Poland. While Mussolini agreed to support Germany militarily, his advisors told him Italy would not be prepared to fight a European power like Britain or France *for several years*. The Germans were told Italy would not be ready for a major European war until *1943*. Hence Mussolini was surprised to learn a few months later of Germany's intention to invade Poland. Mussolini agreed with his son-in-law, foreign minister Galeazzo Ciano, that invasion of Poland would mean a declaration of war by Britain and France, despite assurances from controversial German foreign minister Joachim von Ribbentrop to the contrary.

But Mussolini was right. When Germany invaded Poland on September

1, 1939, Britain and France declared war on Germany. The previous day, Italian foreign minister Ciano assured the British ambassador Italy would not go to war against Britain or France, unless attacked. Moreover, King Victor Emmanuel III, nominally the commander in chief, was aware of Italy's unpreparedness and insisted Italy remain neutral if war should break out between Germany and the Allies—at least until the winner was clear. Mussolini was not given forewarning of Germany's *Blitzkrieg* into France, Holland, and Belgium in 1940. In June of 1940, about a week after the British and French armies were utterly defeated by the Germans and retreated from the continent at Dunkirk, the king relented, and Mussolini finally honored the Pact of Steel. Italy declared war on Britain and France and attacked a few French Alpine outposts and a Riviera town in hopes of spoils from Germany's victory.

Instead, as it turned out, by finally siding with wartime Germany, Mussolini sealed his fate.

In September of 1940, with Britain all but defeated, Mussolini's troops, already fighting the English in East Africa, invaded Egypt from Italy's neighboring North African colony of Libya (whose oil reserves were not discovered until after the War). The goal was to capture the Suez Canal—a vital lifeline for the British who had a treaty with Egypt to defend it—and Alexandria, Britain's major base in the Eastern Mediterranean. Prior to the French and English defeat at Dunkirk, Italy's navy did not dare confront the more powerful British and French navies. Mussolini had called Italy a prisoner of the Mediterranean. With the French fleet now neutralized after France's defeat, and the English defending their homeland against a planned invasion, Mussolini went on the offensive.

Once they conquered Egypt and the Suez, the Italians hoped to acquire Anglo-Egyptian Sudan and control the Red Sea, gateway to the Suez, as well as gain access to oil from the Middle East. However, Italian aggression in both the Red Sea and the Mediterranean failed.

Despite advantages in the number of troops, warships, submarines, and aircraft, the Italians, even with help from the Germans, were never able to dislodge the British from Egypt or Sudan or Malta or Gibraltar. With notable interruptions and despite losses, British sea lanes from the Indian Ocean through the Red Sea to Suez and from there to Gibraltar in the Western Mediterranean remained open, providing Britain with food—it had never been self-sustaining—and vital resources, such as Middle Eastern oil.

Hitler had no prior notice of Italy's October 1940 surprise attack on Greece, which became a disaster—requiring Germany to come to the rescue. On March 25, 1942, Germany, preparing to attack Russia on May 15, signed a treaty with Yugoslavia whereby the Balkan country would join the Axis powers and allow German troops access to Greece. When, two days later, the Yugoslavian government was overthrown in a coup and the treaty recanted, Hitler flew into a violent rage and ordered Yugoslavia attacked. The Axis *Blitzkrieg* invasion of Yugoslavia and Greece, which included a successful, but costly, paratroop assault on the island of Crete, delayed Hitler's invasion of Russia by a critical five weeks in spring of 1941.

Hitler had become the victim of his own success. The relatively quick victories in Poland and France with his *Blitzkrieg* war machine convinced him and his foreign minister, von Ribbentrop, Russia could be conquered in only three months—or so von Ribbentrop told the Italians. In reality, Hitler disregarded the advice of a number of his military advisors that German fuel reserves were inadequate for a prolonged war with the Soviets and that German air power, in light of the losses of aircraft and trained pilots in the Battle of Britain, would be stretched to the limits during such an attack.

Expecting an eventual confrontation with "continental power" the United States, Hitler coveted the resources of another "continental power, the Soviet Union—its fruitful farms, raw materials, and its petroleum from Azerbaijan and the Caucasus. He became convinced he could win a *quick* war. Hitler's racist views colored his thinking.

He believed Slavs were inferior to Germans; and with the pressure of war, the Stalinist regime would collapse from within, as had that of the czar in the First World War. He also had bad intelligence about the numbers and capability of Russian armor, artillery, and aircraft. Finally, Hitler did not prepare for the ability of the Soviets to relocate their defense factories east of the Urals, beyond the range of German bombers, the scorched-earth policy of the Russians in retreating before the attacking Germans, the vulnerability of long supply lines, substantial Anglo-American aid, the fierce tenacity of the Russian people—or the weather.

In short, he gambled on a quick victory and did not prepare for a prolonged war during the Russian winter. Called *Operation Barbarossa,* the biggest invasion in history, Hitler's 3.8-million man army—with over 600,000 horses—launched its attack June 22, 1941.

Mussolini, grateful for the German help in conquering Greece, sent an Italian army of 235,000 men to Russia in support of Germany. The Italians ended up being decimated near Stalingrad. After Italy and Germany were defeated in North Africa and the war turned against Germany on the eastern front, the Italian people turned against Mussolini. By the time he was dismissed as prime minister by the king and held in custody at Gran Sasso in 1943, Mussolini was a broken man. Following his daring rescue from Gran Sasso by German paratroopers, Mussolini told Hitler he wanted to live out his days in peace. Hitler, however, wanted him to become head of the Italian Social Republic to give cover to the German invasion of Italy. Mussolini later said he agreed only because he feared Hitler would otherwise destroy Italy's great cities like Rome, Milan, and Florence.

When Mussolini ruled northern Italy from Salò on Lake Garda, he considered himself almost a prisoner of the Germans. An SS-supervised SD (*Sicherheitsdienst*) detachment watched his every move and monitored his phone calls.

On April 25, 1945, following his frustrating meeting with partisan leaders at the palazzo of Milan's Cardinal Ildefonso Schuster, Mussolini rejected the pleas of many of his followers to stay in Milan. From the Prefecture, or chief government building in Milan, Mussolini gathered documents and money and about 8 p.m. headed north with his German keepers in a convoy containing two armored cars—followed by a number of Fascists in private vehicles—about thirty miles away, to Como, on the southern tip of beautiful Lake Como. The lake, at the foot of the Alps and shaped like the Greek letter λ, was Italy's third largest lake.

Milan was on a war footing. The Allied troops were expected in a few days. Streets were mostly empty. The partisans, many arriving by truck from the countryside, had already seized the newspapers. Although about four thousand members of the Black Brigades, Mussolini's most loyal followers, had gathered at a square in Milan that night, they were only preparing to leave the city. The Fascist era was over.

It was common knowledge the Allies intended to hold show-trials after the war to prosecute senior officials of the Fascist and Nazi regimes. Before leaving Milan, Mussolini had a meeting with his adult son Vittorio at the Prefecture, telling him to remain in Milan and that he feared death less than being tried as a war criminal. He also rejected Vittorio's pleas to fly to neutral Spain.

At about 8:40 p.m., Mussolini and his party arrived at the Prefecture building in Como. Mussolini wanted to remain in Como until his thousands of Black Brigades fighters arrived. However, the Fascist prefect had already been in contact with the local partisans about surrendering, and Mussolini's presence in Como complicated negotiations. Mussolini was urged to try to flee west to Switzerland. Others suggested he go up the eastern side of Lake Como to Sondrio in the Valtellina to make a stand, or go to German-speaking Merano in the South Tyrol region of northern Italy near the Austrian border, new headquarters for his puppet government. Mussolini wanted to await the arrival of a small truck supposedly carrying money and documents from Milan. His son, Vittorio, who ignored his father's advice and fled Milan to hide from the partisans in a monastery near Como, saw the truck, in passing, stalled by the side of the road with its hood up. The truck was never seen again.

Mussolini's private life was complicated. He was named "Benito" by his father after socialist Mexican president *Benito Juarez*. As a young socialist following in his father's footsteps, he did not believe in marriage. He ordered all trace of his early relationship with an Austrian woman named Ida Dalser, with whom he had a son, erased. In a civil ceremony, he married his wife Rachele, the daughter of his widowed innkeeper father's companion, only several years after having a child with her. Altogether, Mussolini had five children with Rachele, two still in their teens. The middle child, Bruno, was killed in an air crash in 1941, flight-testing a heavy bomber. Unlike monogamous, childless Hitler, Mussolini had numerous affairs and fathered an estimated nine illegitimate children by eight different women, a situation to which Rachele turned a blind eye, though she was rumored to have had affairs of her own.

When he left Lake Garda for Milan on April 18, Mussolini told Rachele

Clara "Claretta" Petacci.[c]

Waffen SS 1st Lieutenant Fritz Birzer with Mussolini.[c]

he would soon return, but she never saw him again—though he talked to her by phone and arranged for her to be moved from Lake Garda to a villa near Como. From the Prefecture in Como, Mussolini sent an escort in the early morning hours of April 26 to take Rachele and the two youngest children from the villa, where they were staying, to Switzerland, about twenty miles away. According to Rachele, he sent a note telling her she was his one true love; and if she could not enter Switzerland, she should surrender to the Americans, rather than to the partisans. She was able to get through to him by telephone, and he said his goodbyes over the phone, to her and to the two children with her, urging them to leave. As it happened, the Swiss border was closed (at the request of the Allies) to all fleeing Fascists and Nazis. By the time Rachele and the children returned from their failed trip to the border, Mussolini had left Como. Mussolini's oldest child, daughter Edda, said to be his favorite, was living in Switzerland with three of the sixty-one-year-old dictator's grandchildren, but was estranged from her parents due to Mussolini's approval of the execution for treason of her husband, former foreign minister Galeazzo Ciano, more than a year before.

Mussolini was twenty-eight years older than his long-time mistress Clara—affectionately called "Claretta"—Petacci, who left her villa on the west shore of Lake Garda, near Mussolini's, and followed him to Milan. On April 23, she went to Milan's Linate Airport to see her sister and parents off to neutral Spain. Her father had been a doctor to the prior pope, Pius XI. Her brother Marcello (also a doctor), his consort and children, had remained in Milan. Learning of her lover's sudden departure from Milan to Como the evening of April 25, Clara followed him, with her brother and his family, in a 1939 special-edition Alfa-Romeo Mussolini had given her. She had told several people of her intention to die with Mussolini. At 4:30 a.m. on April 26, Mussolini could wait no longer. He decided to leave Como with three of his generals for Menaggio, about twenty miles north on the west shore of Lake Como. Hearing "Karl Heinz"—the German code name for Mussolini—was about to depart, forty-one-year-old Waffen SS 1st Lieutenant Fritz Birzer pulled his car crosswise, blocking the road. Birzer was in charge of the twenty-to thirty-man German detachment guarding Mussolini, usually commanded by SD Captain Otto Kisnat, who had mysteriously gone back to Lake Garda two days before.

While his men pointed their guns at Mussolini, Birzer insisted Mussolini wait for his German escort. Birzer believed Mussolini had a secret plan to try to escape his German security detachment and enter nearby neutral Switzerland somehow.

With dawn approaching, while much of the rest of the convoy—including a number of Fascist soldiers and two armored cars equipped

Mussolini's armored car (actually a modified Lancia 3 RO truck), operated by a Black Brigades unit from Lucca, with 20mm automatic cannon and three machine guns.[c]

with heavy machine guns—went on to Menaggio, about halfway up the west shore of the lake, Mussolini went to the home of the local Fascist party secretary, on the outskirts of Menaggio, to try to sleep. His top general who was also his defense minister, Rodolfo Graziani, known as the "Butcher of Ethiopia," and two other generals were disgusted at how Mussolini was being misled by Party Secretary Alessandro Pavolini, that *thousands* of Black Brigades troops would defend Mussolini in the Valtellina—*where no preparations had been made.* They abandoned the dictator for a villa that was the SS headquarters in Cernobbio, on the lake, south of Menaggio and just north of Como, site of a drama about to unfold.

Also at the villa was SS General Karl Wolff, key to negotiations for surrender of the German army and SS in Italy. Partisans surrounded the villa. Wolff made desperate phone calls to Switzerland and was eventually rescued by an American OSS team and taken back to Switzerland where he passed through to Austria and then into Northern Italy, to his new headquarters in Bolzano. General Graziani and the two other generals were rescued by OSS Captain Emilio Q. Daddario, who took them to Milan. (Captain Daddario—later to become a congressman from

Connecticut—was able to protect Mussolini's Fascist generals from the partisans and began negotiations for the surrender of whatever German troops remained in Milan).

Clara Petacci arrived in Menaggio in the Alfa-Romeo with her brother Marcello and his family. Marcello Petacci, a friend of the Spanish consul in Milan, hoping to pass himself off as a Spanish diplomat, had put a Spanish flag on the hood of the car and diplomatic tags on the license plates. Clara was taken to the villa where Mussolini had slept for three hours, and the two were reunited.

For unknown reasons, possibly to try to flee to Switzerland, Mussolini and Clara then left in a car, followed that rainy morning by a number of the other Italians, for a small medieval mountain town west of the lake called Grandola, only about five miles from the Swiss border. When SS Lieutenant Birzer had asked where he was going, Mussolini only said, "Follow me," but Birzer's armored car could not keep up.

When Birzer arrived in Grandola, at a hotel taken over by the Fascist militia, he found Mussolini in the lobby, having just walked in the rainy hotel gardens with Clara and three cabinet members.

Clara soon caused a scene.

Mussolini was affectionate toward an attractive twenty-two-year old named Elena Curti, who worked as an aide to Fascist party secretary Alessandro Pavolini; and Clara flew into a jealous rage in the hotel dining room, insisting Mussolini get rid of her.

During the argument, Clara tripped on the carpet and fell, bruising her knee. She then ran into the hotel garden sobbing. Elena actually was widely known to be Mussolini's *daughter* by a prior mistress, a fact later explained to Clara.

Elena offered to go by bicycle to Como to try to find out what was holding up Pavolini and the thousands of Black Brigades fighters from Milan expected to arrive in Menaggio.

Meanwhile, Mussolini received a report that two members of the party he had sent to try to enter nearby Switzerland from Grandola had been arrested by partisans. Learning also of increased partisan activity between Como and Menaggio, Mussolini resolved to return to Menaggio to await his thousands of Black Brigades soldiers, a decision concurred in by SS Lieutenant Birzer and SD Captain Otto Kisnat, Mussolini's regular bodyguard, who had returned from Lake Garda, where Mussolini's government and the German ambassador were being moved from Salò north to Merano in the Alps.

Mussolini's plan was to spend the night at the army barracks in Menaggio and leave before dawn, hopefully with the thousands of troops expected to arrive any time soon. At about 4 a.m. on April 27, Elena Curti—who made the three-hour ride to Como by bicycle the previous rainy afternoon—and her boss, party secretary Pavolini, arrived in Menaggio with a few young soldiers and three armored cars. When Mussolini asked Pavolini how many Black Brigades fighters he brought with him, Pavolini sheepishly responded, "Twelve."

During the night, the Fascist prefect in Como, surrendering to the CLNAI partisans, had made a deal allowing the Black Brigades, which had been subject to partisan attack and Allied strafing, to travel unmolested to a town west of the lake with the intention of surrendering in a day or two to the advancing American 1st Armored Division—then in the process of driving west across the top of the Po Valley and sealing highway exits to the Alps.

All hope of a last stand at the Valtellina was abandoned. Rejected by his countrymen, as he had been in 1943, Mussolini was once again rescued by the Germans—or so it seemed. By chance, a German *Luftwaffe* Flak anti-aircraft unit, headed by 1st Lieutenant Hans Fallmeyer, had arrived in Menaggio from Como during the night. Lieutenant Birzer, now in charge of about thirty SS and SD men and an additional dozen or so German soldiers stationed at the Menaggio barracks, insisted Mussolini and his party join the German convoy, which set out from Menaggio at 5:30 a.m. on April 27, supposedly headed for Merano in the German-speaking part of northern Italy, near the Austrian border and about twenty miles from Bolzano. Birzer later said he believed the convoy had about three hundred people, between the Germans and Italians, including the civilians. Mussolini was confident the Germans would protect him, his cabinet ministers and their families, and other Fascists in the convoy as they traveled the narrow, two-lane mountain road, called *Via Regina*, north around the rest of the west shore of Lake Como.

At first Mussolini rode in a camouflaged car; but upon reports of partisan activity in the area, he signaled for the convoy to stop and moved into the armored car that was in the lead and carried several top Fascists, like party secretary and Black Brigades head Pavolini, (as well as Elena Curti, Mussolini's secret daughter). The armored car was like a rubber-tired tank, with a 20mm cannon in the turret on top. Clara Petacci was somewhere behind in the Alfa-Romeo with her brother, who posed as a Spanish diplomat, and his family.

At about 7 a.m., after the convoy had passed through the village of Musso and was on the northern outskirts of the town, the right rear tire of Mussolini's armored car was punctured by a three-nail partisan booby trap. At almost the same time on the narrow mountain road overlooking the lake between Musso and Dongo—with a high rock retaining-wall on the left and three-foot rock safety-wall, with a steep drop, on the right—the armored car encountered a partisan roadblock consisting of a large tree trunk and boulders. A warning machine-gun burst was fired and a red partisan flag went up.

Return machine-gun fire from Mussolini's armored car killed an old man—not a partisan—on his way to work. Then the shooting stopped. According to one of the partisans who was there, two German officers preceded by two soldiers waving white flags approached the roadblock

Mussolini with Elena Curti (his secret daughter), who worked for party secretary and head of Black Brigades Alessandro Pavolini (extreme left with moustache). At Milan Prefecture. April 25, 1945.c

and two men with red scarves around their necks—a sign they were Communist partisans—came out. The Germans were asked to wait a few minutes while the partisan commanding officer was summoned from the next town, Dongo, only a half-mile away. A short time later, the Germans started negotiating with a brash twenty-four-year-old non-Communist who, nevertheless, had been named, a month earlier, commander of the Puecher Detachment,52nd Garibaldi Brigade "Luigi Clerici," a Communist unit that had come out of the mountains a day or two previously and now had control of the northern third of the west shore of the lake. He was of Florentine nobility, had been raised in Tuscan Pistoia, and went by the partisan name "Pedro." His real name was Count Pier Luigi Bellini delle Stelle. He was to be given credit as the man in charge of capturing Mussolini—but that happened later.

Lieutenant Fallmeyer, the German anti-aircraft officer who was in overall command of the convoy, spoke good Italian and apparently had talked his way through partisan roadblocks en route to Mennagio. He told delle Stelle he had orders to take his anti-aircraft unit to Merano and had no intention of fighting the Italians. Furthermore, he insisted the partisan leadership had made an agreement with the German High Command that German troops were to be allowed through partisan territory without confrontation. Of course, no such agreement had been formally made, for it would have violated the plan of the Allies—who had armed and supported the partisans—to prevent the Germans from fleeing Italy. On the other hand, Lieutenant Fallmeyer's troops had been allowed through previous roadblocks by partisans who did not want to take on a heavily-armed German unit that only wanted to avoid confrontation.

Count delle Stelle put up his own bluff. He said his mortars and machine guns could destroy the convoy in fifteen minutes, and he had no orders to let the Germans through. In actuality, he had fewer than twenty men, most armed with shotguns, and fewer than five had stopped the convoy. A fellow partisan whispered to him the Germans had at least twenty-eight open and covered trucks and armored cars, each armed with a heavy machine gun, plus mortars, at least two hundred soldiers, many with submachine guns and grenades, and several pieces of towed anti-aircraft artillery.

The Germans had a protestant chaplain along. As citizens of Musso

Count delle Stelle "Pedro."

came to the road to see what was happening, the chaplain spotted a catholic priest among them, Father Mainetti, and approached him, hoping the priest would have influence with the partisans and try to avoid bloodshed. The chaplain did not speak Italian but had taken some Latin and asked the priest —who, of course also knew Latin—how many partisans were in the area. For "partisan" he probably used Latin words for "citizen-soldier." In any event, the priest thought he was being asked how many citizens lived in the area, and he responded, in Latin, "About three thousand." While Lieutenant Fallmeyer and the other German officers were discussing what to do, the chaplain erroneously informed them he had heard from the priest the partisan force was *three thousand* strong—a much larger force than Fallmeyer thought he was facing.

Count delle Stelle told Lieutenant Fallmeyer, even if the convoy got through the roadblock, the partisans had mined a bridge ahead. It wasn't true—at least not yet. Stalling for time, and figuring the convoy would not go anywhere without its leader, delle Stelle suggested Lieutenant Fallmeyer accompany him in a German car with a German driver to partisan headquarters in the next sector, so delle Stelle could get instructions about what to do. It was about 8:00 a.m. and a radio broadcast from Milan, picked up by vehicles in the convoy, reported the false claim Mussolini had been captured elsewhere.

Delle Stelle took the German commander on a bit of a *goose chase*. They went through several roadblocks between Dongo and the north end of the lake. Delle Stelle left word for his men to be prominent along the roadway and to set up various machine gun emplacements with weapons captured from surrendered German and Black Brigades garrisons the day before. He wanted Lieutenant Fallmeyer to surrender instead of fight, and the greater the partisan show of force, the better would be delle Stelle's negotiating position. Delle Stelle learned the partisan chief north of Lake Como was in a town called Chiavenna. At the north end of the lake, he left Lieutenant Fallmeyer with one of his men in Nuova Olonia, at a surrendered German garrison, and requisitioned a car. At Chiavenna, he was told to deal with the situation as best he could, but at all costs to prevent the German column from coming to Chiavenna armed. The partisans there were fighting some holdout Germans and Fascists. Delle Stelle returned to Lieutenant Fallmeyer and told him he would offer his terms when they got back to Dongo, which he did.

The Germans were told they could keep their arms for the time being and go through delle Stelle's sector to the northernmost point of the lake. There, they would be subject to another roadblock and would await further authorization before entering the next partisan sector—*but only after submitting to an inspection in Dongo.* All Italians traveling in the convoy, however, would be detained until their identities were established. Fallmeyer said he could not agree until after he spoke to his fellow German officers and to the Italians in the convoy.

As hour after hour had gone by back in Musso, the soldiers and civilians had grown more and more impatient. Several of the Fascist officials left their cars at the tail end of the convoy around lunch time and visited the priest, Father Mainetti, at his rectory, by the church, asking for refuge or to request he look after family members. The priest offered the various cabinet ministers soup and was soon aware from their conversations among themselves Mussolini was in the convoy. The priest sent a messenger to tell the partisans not to let Mussolini escape.

Lieutenant Fallmeyer did not return until about 2 p.m., having been away for over six hours, and discussed the situation with the other officers, including SS Lieutenant Birzer and Otto Kisnat, an SD captain (who nevertheless was outranked by Birzer of the SS). Fallmeyer had dropped delle Stelle off in Dongo on the way back. When delle Stelle learned of the presence in the convoy of the hierarchy of the Fascist government and party—and possibly Mussolini himself—delle Stelle sent ten extra partisans back to Musso to make sure no one escaped.

When delle Stelle returned to Musso, a short time later, Lieutenant Fallmeyer told him the Germans accepted his terms and would submit to an inspection in Dongo. Delle Stelle told several of the Italians in the group the Italians could not follow the Germans to Dongo.

Before delle Stelle had arrived, SS Lieutenant Birzer suggested Mussolini be removed from the armored car and be put in one of the German trucks dressed like a German soldier. Mussolini was reluctant

Lake Como area north of Milan.

to do so, but Clara Petacci had come up and told him to try to save himself. Mussolini supposedly said at that point he trusted the Germans more than his Italian countrymen. Several suitcases with documents and money brought from Lake Garda were also put on the truck.

About 3 p.m., the armored car, with the tire now repaired, moved aside, and all the Germans, including Mussolini's security detachment, were allowed through the roadblock and headed for the main square, or piazza, in Dongo by the lakeshore, to be inspected by the partisans. Behind the Germans was the Alfa-Romeo with the *supposedly* Spanish consul (really Marcello Petacci), his consort Zita Ritossa—they never married—their two sons, ages three and six, and sister Clara in her mink coat.

Back in Musso, the military officers, party leaders, cabinet ministers, and others in Mussolini's government felt betrayed. They numbered about fifty, plus spouses and children. Fascist Francesco Barracu, a decorated World War I veteran, negotiated with delle Stelle about what to do with the armored car which had led the convoy and was at the side of the road to let the Germans through. Barracu insisted he and its other occupants wanted to drive it to the far-off Adriatic port of Trieste, over three hundred miles away, won from Austria in World War I, to help prevent Trieste from being overtaken by the Communist Yugoslavs. Delle Stelle told him that made no sense, and besides, they would never make it through all the roadblocks on the way. He refused to allow the armored car to go north to Dongo where the German trucks were being searched, three at a time. Wanting to make sure Mussolini was not inside, delle Stelle ordered the rear door opened. He looked inside and recognized no one. Reluctantly he gave permission for the vehicle to turn around in fifteen minutes and head back to Como; but the partisans were to shoot if it tried to go to Dongo.

When the armored car, containing party secretary Alessandro Pavolini, Elena Curti, and about twenty others, moved forward to turn around, the partisans, believing it was heading for Dongo, opened fire with a machine gun and threw a hand grenade under it. The armored car returned fire, wounding two partisans; but a bullet killed a Fascist inside, and a white flag was soon waved. Pavolini and six others ran from the rear door of the vehicle toward a precipice and the lake, some distance away. Another Fascist was killed outside the vehicle. Only Pavolini and one other official made it to the lake without being caught. Partisans in a boat found them in shallow water, hiding under a rocky overhang, about an hour later. When Pavolini tried to shoot at them, a partisan fired a shotgun at him from the boat, wounding him slightly in the upper torso. The two cold, wet fugitives were taken by boat to Dongo, where Mussolini already had been arrested.

Pavolini was especially despised by the partisans.

The paramilitary Black Brigades—harkening back to the "Blackshirts," who originally brought Mussolini to power—were organized by Pavolini for Mussolini's Salò Republic with the primary purpose of hunting down the partisan fighters and their civilian collaborators.

Up until two days previously, a Fascist Black Brigades detachment had worked out of the town hall in Dongo and angered the townspeople by killing four partisans and leaving their bodies to rot in the countryside. An illegal strike was called at the Falck steelworks factory in Dongo so the enraged workers could search for the bodies and give their heroes a proper burial.

While Pavolini was being captured, the partisans were inspecting the German vehicles in the rainy Dongo town square, on the lakeshore, with a number of angry townspeople looking on. The people of Dongo hated the Germans almost as much as they hated the Black Brigades.

Neither the Germans, nor the partisans, nor the townspeople had any idea of the *carnage* that would occur there the following day.

The partisan who inspected the fourth or fifth truck in the convoy, apparently open, could see a soldier seated near the cab, with a blanket over him, between two gasoline drums, who was out of place. His German Flak helmet did not seem to fit right. He was in sunglasses and wearing a German great coat with the collar up. What gave him away was his riding boots and the black stripes on his trousers, unlike what the other Germans were wearing. The soldiers on the truck insisted the man was drunk. The partisan suddenly recognized the most famous face in Italy—but, fearful of a confrontation, pretended not to. Instead, he reported his find *in dialect*—the way the partisans communicated with each other—to delle Stelle's second-in-command, twenty-year old Urbano Lazzaro, who at the time was inspecting another truck and went by the partisan name "Bill."

After first trying to rouse Mussolini from outside the truck full of seated Germans, and calling him by name, Lazzaro fearlessly climbed aboard at the rear and approached the supposed drunk. Mussolini was seated with the stock of a submachine gun under his chin. Lazzaro took the weapon from him, removed the helmet, revealing Mussolini's bald head, took off the sunglasses, and folded down the collar of the great coat. He asked if Mussolini had any other weapons. The captive unbuttoned his coat, reached into his waistband, and produced a 9mm Glisenti automatic pistol with one butt-plate missing. Lazzaro put the weapon in his pocket, told Mussolini he was arresting him in the name of the Italian people, and helped him to stand. Mussolini motioned to the Germans not to defend him. Lazzaro told his prisoner not to worry, that he would not harm a hair on his head—later realizing the irony: *Mussolini had a shaved head.*

The tailgate was lowered; and the prisoner was helped off the truck and taken through the large crowd of derisive onlookers to a room in the town hall overlooking the square. He was soon joined by other arrested members of his party, eventually including the wet and wounded Pavolini, who saluted him.

After the search, the German convoy was allowed to leave Dongo and go to the north end of the lake to await further authorization to cross a bridge and proceed through more partisan territory. Meanwhile, Marcello Petacci's Spanish documents were pronounced fakes by Lazzaro; and Marcello and his family were taken under guard to the Hotel Dongo, near the square. Clara Petacci, having no identifying documents and claiming she had simply been given a ride, was arrested and taken to a room on the ground floor of the town hall. Lazzaro went back to Musso to inform delle Stelle of Mussolini's capture. The partisans there were still in the process of trying to identify all the Italians who were in the convoy.

Lazzaro and delle Stelle returned to Dongo, to the town hall. Lazzaro, a former Finance Guard official, went through Mussolini's briefcase, cataloguing the documents inside and making a record of the various bank drafts, worth nearly two million lire, and 160 gold coins. He then proceeded to the roadblock at the north end of the lake, stopping in a town called Domaso to deposit in a savings bank Mussolini's briefcase, and that of another prisoner. He had not discovered the other cases Mussolini brought with him on the German truck.

The Mera River runs into the north end of Lake Como, and all traffic must cross a bridge over the river at Ponte del Passo, where the Germans had been halted by a partisan roadblock. When Lazzaro arrived, he ordered the Germans to surrender and *disarm*. He said the bridge ahead was mined—it was not. He also warned if the bridge were blown the convoy would be stuck and in range of a gun emplacement across the lake—it was actually still manned by Germans. Lieutenant Fallmeyer met with his officers and agreed to surrender, but only *in writing*, and to a senior partisan officer. As it was getting dark, he was led to the area commander, headquartered in a school in Morbegno, about twelve miles east, and a formal agreement was drawn up and signed—effective the following morning. The Germans agreed to be disarmed at Morbegno, after which they could travel through partisan territory to Switzerland, or elsewhere.

On the road back to Domaso, Lazzaro encountered a Fiat car carrying a German captain and a second lieutenant. They said they were trying to catch up with the convoy, which had some of their luggage. Lazzaro appropriated their car and sent a German-speaking associate to drive the two Germans to get their luggage and bring them to Domaso. Later that night, the associate said the Germans gave him over 33 million lire, 11 million for the associate, and the rest supposedly to be split between two mistresses left behind in Tuscany. The money was deposited in the Domaso bank.

The German convoy, after camping for the night at Ponte del Passo and dumping or burning the cases Mussolini left on the truck, was disarmed the next morning in Morbegno, went by way of Chiavenna to the Swiss border and was allowed to cross, after a long wait—the only German unit allowed to do so—and go northeast through the Maloja Pass and Alpine Switzerland's Engadin/Inn River Valley through St. Moritz to the Austrian border near Nauders and the Resia (Reschen) Pass near German-occupied Italy. (Had Mussolini made it past Dongo, he might have made it to Switzerland or Austria).

The partisans around Lake Como had a powerful ally in the militarized Italian police force called the *Guardia di Finanza*, or **Finance Guard**, responsible for enforcing tax laws, immigration, counterfeiting, white-collar crime, customs—just about anything having to do with money. When Italy switched sides in 1943, the Finance Guard was so respected it was allowed to continue to operate in the interest of the Italian people not only in the areas under Allied control, but even in enemy territory. So, the

Germans allowed the Finance Guard to operate in Northern Italy, in theory as part of Mussolini's Italian Social Republic.

However, starting in 1945, the head of the Finance Guard for Lombardy, headquartered in Milan, started working more and more with the partisans. When Mussolini left Milan the evening of April 25, the uniformed Finance Guard helped keep public order while many of the partisan fighters, still mostly in the hills and rural areas, were yet to arrive in the city. The Finance Guard also arrested fleeing Fascists at border checkpoints and often helped partisan fighters in the hills get into and out of the Italian-speaking Swiss canton of Ticino, which bordered the Lake Como area.

Count delle Stelle was wary about transferring Mussolini from the Dongo town hall to the jail for fear the dictator would be freed by the Germans or Fascists—or killed by angry local people. In the late afternoon on April 27, he and a Finance Guard sergeant took Mussolini to a more secluded place, the Finance Guard barracks in Germasino, which had a jail cell and was about three miles from Dongo and about 2,000 feet up in the mountains.

Dongo was a factory town on the lake at the mouth of the Albano Creek. Germasino, near iron ore deposits in the mountains, was at the upper end of the Albano Valley, past the Falck steelworks, which employed 1,500, had been in the valley over a hundred years, and had been converted by the Germans and Fascists into wartime production, making parts for mortars and *Panzerfaust* anti-tank weapons.

When delle Stelle was about to go back down to Dongo, Mussolini asked him to give his regards to the lady traveling with the Spanish gentleman. The charming delle Stelle asked why and got Mussolini to reveal the lady's identity. Back in Dongo, delle Stelle arranged for several more prisoners to be taken up to Germasino. Others were given straw mats to sleep on in the town hall. He went to the town hall room where Clara Petacci was kept and talked to her for nearly an hour. Finally, she admitted who she was. She asked to be taken to Mussolini and said she wanted to die with him.

Up the chain of command, delle Stelle sent a message to the partisan commander in Como asking what he should do with the captured Mussolini and hierarchy of the Fascist government. Word was also passed on to Milan, where Luigi Longo, top Communist in the partisan military arm, received a telegram at about 8:30 p.m. that Mussolini, Pavolini, and others were in the custody of the Puecher detachment of the 52nd Garibaldi Brigade. Another telegram came about two hours later with more specifics of who was in custody.

The partisan leader in Como, fearful of the arrival of the Allies there at any time, made arrangements for Mussolini to stay in a safe house on the east side of the lake, outside of his city. He left word for delle Stelle to disguise Mussolini with bandages and bring him in the early morning hours to Moltrasio on the west shore of the lake. From there, a boat was to take the prisoner across the lake to the safe house, thus avoiding Como, at the very south end of the lake. If anyone asked, the prisoner was a wounded partisan.

Shortly after 1 a.m., on April 28, delle Stelle was in a car with a bandaged Mussolini—the only parts of his head that were visible were his eyes and mouth. Mussolini had refused to wear the German great coat he wore when arrested and had been given a Finance Guard overcoat. Also in the car was a female partisan only recently freed from Fascist imprisonment and torture, who was to pretend to be a nurse. At a designated spot, delle Stelle's car met with a second car that contained several more partisans, one well known in the southern half of the lake, and Clara Petacci, who also was supposed to act as a nurse. When the two cars reached the rendezvous, Clara and Mussolini were reunited briefly, but then continued on in separate cars.

The rain and fog around the lake made driving difficult. The partisans stopped by a hotel in Moltrasio, but could not find the boat. In the distance, they saw rockets over Como and speculated advance elements of the Allied Armies had arrived there. Delle Stelle was aware the Allies wanted Mussolini for themselves so the dictator could be tried for war crimes. However, like the other partisans, delle Stelle did not want Mussolini to fall into Allied hands. He knew the CLNAI partisan headquarters was supposed to turn Mussolini over to the Allies; but he felt the Italian people should deal with Mussolini—exactly how was not up to him.

With the original plan now too risky, delle Stelle ordered the two cars to take Mussolini and Petacci back north, about ten miles away, to a farm house, known to one of the partisans, on a hill overlooking the lakeside town of Azzano. The three-story house, technically in a hamlet called *Bonzanigo,* was owned by a kindly, trustworthy couple named De Maria, who were delighted to put up the prisoners in a bedroom normally used by their two sons.

At about 5 a.m. three of the partisans headed off to Como in one car, to tell party officials there what happened, while delle Stelle and the rest left for Dongo in the other car, leaving two partisans—fishermen, actually—behind to guard the exhausted Mussolini and Petacci.

Delle Stelle never saw the two prisoners again.

40

THE DEATH OF MUSSOLINI

To some extent, the events of April 28–29, 1945 began the previous August when a German truck distributing food on a city street to hungry citizens of Milan was bombed by partisans. The six dead were all civilians. No Germans were killed, and only the driver was slightly wounded. Nevertheless, in retribution, the Germans ordered removal of fifteen anti-Nazi and anti-Fascist prisoners—mostly Communists and Socialists—from Milan's notorious *San Vittore* prison. A firing squad of Fascist militiamen, under orders from the German SS, executed the fifteen prisoners at dawn on August 10, 1944 in a public square in Milan called the *Piazzale Loreto*. The dead bodies remained on display on a walkway all day in the hot August sun, supposedly as a ghastly warning to the people of Milan.

Ironically, even Mussolini, when he learned of the massacre, disapproved, saying—prophetically, as it turned out—the Fascists would likely pay a high price for it. The top Fascist official in Milan was so disgusted he resigned over the incident. In memory of the massacre, Piazzale Loreto was thereafter called by many "*Piazza Quindici Martiri*" (Piazza of Fifteen Martyrs).

In the early afternoon of Saturday, April 28, still operating on little sleep, delle Stelle received a message in his newly acquired office in the Dongo town hall that a large, covered, yellow truck, full of armed men, and a black car had gone through a checkpoint and were headed toward Dongo. As delle Stelle received the message, a partisan entered to tell him the truck and car were already in the piazza in front of the town hall, and the leader of the armed men wanted to see the local commander. At first delle Stelle feared a Fascist rescue attempt. The messenger was to reply that delle Stelle was available *in his office*. A few seconds later, the breathless partisan returned to report the man was from Milan and wanted delle Stelle to come to him *outside* immediately. So, delle Stelle met the tall, dark, impatient man in his mid-thirties, who called himself *Colonel Valerio* and invited him inside. At delle Stelle's suggestion, the armed men from the truck, all in new uniforms—not like his ragged partisans—went to eat, while delle Stelle, now fairly certain he was dealing with a senior partisan, took Valerio to his office. Valerio sat down behind delle Stelle's desk, as if to show he was now in charge, and presented his credentials.

Meanwhile, another car arrived in the town square. In the car were two Communist party officials from Como, plus partisan *Riccardo*, commander of the militiamen from Milan, and *Guido*, co-leader of the expedition who had left Valerio at the Prefecture, or government building, in Como to meet separately with Communist leaders, at the new Como party headquarters—only to learn later Colonel Valerio had left for Dongo without him.

Valerio berated his co-leader Guido for having left him in Como with no explanation. He actually suspected Guido was trying to beat him to Dongo so he could claim the glory of killing Mussolini himself. Guido tried to cool Valerio's temper saying he needed Como partisans to ease the way with the detachment of the 52nd Garibaldi Brigade in Dongo. One of the men who accompanied him to Dongo had met the Communist party commissar for delle Stelle's detachment, *Pietro Gatti* (Michele Moretti), early that morning at the old party headquarters in Como, where the tired partisan had gone from the safe house in Azzano to explain why Mussolini had not been turned over to Como control. In Dongo, Pietro Gatti was able to vouch for the Como official, who vouched for mission co-leaders Guido and Valerio, which helped somewhat take the chill off what initially had been an icy Dongo reception for the big-shot Communist partisans from Milan.

The previous night, Luigi Longo, top Communist military leader in Milan, after hearing of the capture of Mussolini and the Fascist hierarchy, ordered the gruff "Valerio"—only years later, when running for public office, identified as *Walter Audisio*—and the older and more level-headed *Aldo Lampredi*—"Guido"—to set out first thing in the morning from Milan for Como and Dongo with a platoon of men (twelve to fourteen) armed with submachine guns and grenades, under command of *Alfredo Mordini*—"Riccardo"—to execute Mussolini and fifteen of the Fascist hierarchy. The false reports of Mussolini's capture earlier in the day had prompted Allied officials in Sienna to wire a reminder to the partisan government in Milan that Mussolini, if captured, was to be turned over to the Allies. The partisan government was aware the Allied troops would be entering Milan any day to impose military government, and would demand custody of Mussolini, if in captivity. Audisio and Lampredi were to pretend they were supposed to convey Mussolini to Milan. Their real mission was to kill Mussolini before the Allies got to Milan, and to avenge the fifteen executed in the Piazzale Loreto massacre the previous August.

With advance elements of the 1st Armored Division now in nearby Como, Audisio and Lampredi would need permission from the highest source to go through any military roadblock that might prevent their going to Dongo and returning, so a pass was obtained during the night through the office of the top military commander of the partisans, Raffaele Cadorna, signed by American OSS Captain Emilio Q. Daddario, who had just arrived in Milan after rescuing Mussolini's generals from the partisans. Audisio said his mission was to bring Mussolini to Milan. The pass was in English with an Italian translation. Audisio had gotten a late start because he could not find an appropriate truck for his men, first in Milan, where he could only get a small, run-down pickup, and then in Como where he and Lampredi arrived with their men a little after 8 a.m.

In Como, at the prefecture, or chief government building, all was in chaos as CLNAI partisans were in the process of taking over from the Fascists, who had surrendered. Mussolini was supposed to have been brought to Como in bandages as a disguise early in the morning, but had not come. Audisio was offered an ambulance, not an acceptable truck for his execution squad. Hours went by. Audisio put in a call to his boss Luigi Longo in Milan. When Longo called back, he told Audisio to find a way to take out Mussolini—"or we will take you out." Meanwhile, with time passing, the hot-headed Audisio learned Lampredi and Mordini had *disappeared* with the car they had brought from Milan. (They had gone to the new local party headquarters, taken over from the Fascists). Finally, after an angry altercation with the politicians in Como, Audisio and his men abruptly left in requisitioned cars at around noon. The president of the local council and the military commander hurriedly got into a military car and accompanied him to make sure Mussolini was brought to Como. Off the street outside Como, Audisio commandeered the large Fiat 634 covered truck he brought to Dongo (said to have been for hauling furniture). Once in Dongo, he had the two officials from Como placed under guard and told delle Stelle he was behind schedule and intended to shoot Mussolini and fifteen of the arrested Fascist leaders—the same number as had been executed in Milan's Piazzale Loreto the previous August. He asked delle Stelle for a list of the prisoners.

As Audisio was putting crosses by the names of those he chose to be executed, delle Stelle—who practiced law after the war—tried to intercede on behalf of a few, such as Clara Petacci, saying they were not deserving of death, at least not without a trial. Audisio, however, would have none of it. The partisan central government had already condemned all the fascist hierarchy to death, though not by name, and Audisio insisted he alone had authority to decide who would be among the fifteen to be executed. He branded Petacci an "advisor" to Mussolini and worthy of death. An unknown was the so-called Spanish consul—really Marcello Petacci, Clara's brother—whose papers appeared forged and who spoke Italian like a native. Audisio, who had fought against Franco in the Spanish civil war and knew Castilian, questioned Marcello Petacci, who insisted he was from Spain—but obviously spoke no Spanish. Audisio was convinced the man was Mussolini's oldest son, Vittorio, and ordered him to remove his shoes and be immediately taken to be executed.

Audisio wanted all the prisoners brought to the lakeside piazza in Dongo. Delle Stelle informed him while some of the prisoners were in Dongo, others were in the Finance Guard barracks in Germasino; and Mussolini and Petacci were being guarded at the De Maria safe house overlooking Azzano. Delle Stelle told Audisio he would bring the prisoners back from Germasino; and two of his men, *Michele Moretti*—the aforementioned "Pietro Gatti"—and *Luigi Canali*—"Neri"—who were involved in taking Mussolini and Petacci to the safe house, would bring them back to Dongo.

Once delle Stelle left for Germasino to get the prisoners there, Audisio and Lampredi, leaders of the Milan group, decided to accompany delle Stelle's men to get Mussolini and Petacci. While delle Stelle's man Moretti was a committed Communist, they did not trust Canali, and left him behind. A black Fiat 1100 and driver were requisitioned on the spot; and Audisio, Lampredi, and Moretti left Dongo. They had no intention of bringing Mussolini and Petacci back. Instead, a few blocks from the safe house, they chose an unassuming quiet spot on Via XXIV Maggio to do the deed. They also resolved to pretend they were rescuing Mussolini and Petacci to gain their cooperation.

The safe house, elevated from the street, had no driveway and was only accessible up a long, paving-stone walkway with several steps. The armed executioners—Audisio had a pistol and he and Moretti submachine guns, and Lampredi a pistol—left the car and driver as close as possible. The two fishermen guarding the bedroom door—shoeless in deference to the De Marias, their hosts—recognized Moretti. When the bedroom was entered, Mussolini, standing on the right, an eyeglass case in each hand, raised his arms. Petacci, though dressed, was in bed. Audisio told them he was there to free them, but they must hurry. Mussolini put on the Finance Guard jacket he had brought from Germasino and was asked to wear a beret, so as not to be as recognizable. Petacci put on her mink coat. Menstruating, she was not wearing her panties and was looking for them in the bedding when Audisio told her she had to move quickly. She left behind the panties and her purse in the rush to get out the door. She had difficulty in her high heels going down the walkway to the car.

Mussolini appeared confident he was being rescued. Clara was more suspicious.

With Mussolini and Petacci in the back seat, Lampredi in the front passenger seat next to the driver—identified many years later as Giovanni Battista Geninazza—and Audisio and Moretti sitting on the front fenders,

Audisio facing backwards, not letting Mussolini out of his sight—the car very slowly coasted the short distance downhill (about a quarter mile) to the spot previously chosen, right after a turn in the road—the gate of the Villa Belmonte, **Via XXIV Maggio, No. 14**, technically in the town of *Giulino di Mezzegra.*

Audisio, pretending he heard a suspicious sound, ordered the car to stop.

Mussolini and Petacci, somewhat bewildered, were ordered out of the car and made to stand to the left of the gate in front of a short stone wall.

What happened next was a bit of a *comedy of errors.*

Audisio aimed at Mussolini the Thompson submachine gun he had taken from the arms locker in his office in Milan early that morning and announced he was executing Mussolini in the name of the Italian people. Petacci clung to Mussolini. Audisio told her to get out of the way, or she would be shot, too. Mussolini seemed dumbfounded, disbelieving his "rescuers" were really *executioners.* When Petacci moved to Mussolini's right, Audisio, still aiming at Mussolini, pulled the trigger.

Click.

The weapon jammed. He manipulated the bolt handle, re-aimed, and pulled the trigger again.

Click.

Lampredi, to the right of Audisio, pulled a pistol, apparently a 9mm Beretta, from the pocket of his trench-coat. He supposedly aimed it at Mussolini, and pulled the trigger. (Audisio claimed he took out his own pistol, aimed it at Mussolini, and pulled the trigger). Under either scenario . . .

Click. It too jammed.

Mussolini, who had survived at least four assassination attempts, and seemed to lead a charmed life, still appeared confused. With Lampredi now holding both useless guns, Audisio called Moretti over and demanded Moretti's weapon, a French-made 1938 MAS 7.65mm submachine gun delle Stelle had given Moretti after it was confiscated in a partisan raid on a Black Brigades barracks in Gravedona a month before.

Petacci got hysterical, insisting they could not shoot Mussolini, and embraced her companion. Finally, comprehending the situation, Mussolini pushed her out of the line of fire and opened his jacket asking to be shot in the heart. Audisio fired a short burst.

Michele Moretti *Pietro Gatti*.[c]

Walter Audisio *Colonel Valerio*.[c]

Aldo Lampredi *Guido*.[c]

As Mussolini was falling, Petacci jumped in front of him, and Audisio fired a burst at her, as well. Mussolini was still alive, however, and Audisio supposedly fired a single round from the MAS (some say from a pistol) at point-blank range into Mussolini's heart. The time was about 4:10 p.m.

Moretti then picked up the shell casings. The two fishermen, who finally had their shoes on, had taken a different route. Hearing the gunshots, they arrived after the shooting and were ordered to guard the bodies, while Audisio, Lampredi, and Moretti headed back to Dongo, thirteen miles away, in the car.

With some variations, that is the generally accepted version of how Mussolini was killed. There were only four witnesses—the three Communist partisans and the driver. The true identities of the participants

Farmhouse of Giacomo and Lia De Maria where Mussolini and Petacci were put up.[c]

The driver, Giovanni Battista Geninazza, with his Fiat 1100.[c]

were not revealed for several years, for fear of retaliation from hard-line Fascist sympathizers.

The matter also had political overtones. The first versions of what happened, from Audisio, using only the partisan names—Valerio, Guido and Pietro Gatti—were published in the Communist newspaper *L'Unità* in 1945, and another version in 1947. Lampredi gave no interviews but submitted his version to the party in 1972. However, speculation continues to

GoogleMaps view of place where Mussolini and Petacci were executed. Black cross on wall next to left gate pilaster marks the spot with inscription: "Benito Mussolini 28 Aprile, 1945." (©Google 2017).

this day. Was Moretti the actual shooter while Audisio took credit? Conspiracy theories arose. Did Winston Churchill order the assassination to save himself the embarrassment of exposing communications with Mussolini behind the back of the Russians and Americans? Did the execution take place earlier in the day under a British agent, with the events on Via XXIV Maggio staged? Did communist official Luigi Longo sneak out of Milan and do the shooting?

Even fifty or sixty years later, new books and articles were still coming out claiming to tell the *true story* of how Mussolini died.

41

THE DONGO MASSACRE

Delle Stelle put all the chosen prisoners together, including the ones from the Finance Guard barracks in Germasino, in the ornate, neoclassical *Sala d'Oro*, or "Golden Room" of Dongo's town hall, once the mansion of a wealthy family. He then went down to the building's entrance, off the piazza, where he encountered an excited Audisio, just back from Azzano. "Justice is done!" Audisio said. "Mussolini is dead!" Delle Stelle was shocked. Audisio was expected to remain in Dongo while delle Stelle's men, Moretti and Canali, got Mussolini and Petacci. Audisio waved his hand. "I know, I know," he said. "But I couldn't waste any more time. It was late, and I had already lost several precious hours."

Audisio was too preoccupied with making arrangements for the executions in Dongo to go into detail about how he had killed Mussolini and Petacci; but Moretti excitedly showed delle Stelle his MAS submachine gun, the one delle Stelle had given him, and said, "This is the weapon that killed the tyrant!"

Delle Stelle led Audisio up to the "Golden Room" for final decisions

The Golden Room.

on the fifteen to be executed. On a stair landing, they encountered Giuseppe Rubini, named the CLNAI mayor of Dongo two days earlier when the Black Brigade detachment departed. He stated his objections to having the executions done in his town—and especially in the lakeside public square in front of the town hall, where women and children would be present—and suggested several *more-secluded* places. Audisio said he had orders to hold a *public* execution, and refused to change the location. Rubini said he would resign as mayor so as to avoid all responsibility.

In the "Golden Room," Audisio finalized his choices. He made a mistake. A Fascist Air Force fighter pilot named Calistri was believed by Audisio to be Mussolini's personal pilot. Instead, the stranded officer had joined Lieutenant Fallmeyer's convoy to get out of Como. Some minor officials were weeded out. The fifteen condemned to die were made to stand on one side of the room. A friar, Father Accursio Ferrari from

Franciscan priest Fr. Ferrari with the prisoners in the Dongo piazza.

the local monastery, wanted to meet with the prisoners one-on-one in the town hall for the sacrament of Confession and last rites. Audisio said *no*, but he would allow the priest three minutes for general absolution just before the prisoners were shot. A partisan approached Audisio and said the man claiming to be the Spanish consul insisted on having his papers re-examined and wanted to be interrogated again. Audisio sent the partisan away.

Delle Stelle wanted nothing to do with the actual execution and said he would not attend. The plan was for one of delle Stelle's men to accompany each of the fifteen condemned men from the town hall and across the piazza, where custody of the prisoners officially would be turned over to Mordini, head of the execution platoon that had come from Milan. Audisio ordered delle Stelle to attend so he could witness everything was done properly. (Afterwards, delle Stelle insisted his men did not participate in the actual execution, whereas Audisio said several of delle Stelle's men were in the firing squad, which was to have one member with a submachine gun a few yards opposite each of the fifteen prisoners).

When the prisoners were lined up at the railing at the end of the piazza, overlooking the lake, Father Ferrari was allowed his few minutes. While the priest was trying to give words of comfort, absolving the condemned men of their sins, Audisio remembered the Spanish consul and asked delle Stelle why he had not been executed earlier in the day, as ordered. Delle Stelle told him the man was not Vittorio Mussolini and not the Spanish consul, but rather Clara Petacci's brother. However, the other prisoners seemed to despise him and would not identify him as Marcello Petacci. (The Fascists viewed the Petaccis as profiteers who took advantage of Clara's relationship with Mussolini). Audisio ordered

Marcello Petacci to be brought out and shot with the others. When the priest was finished and Petacci was taken to the lineup of prisoners, the men about to be shot objected, saying Petacci was a traitor, not a true Fascist, and not worthy to be executed with them. Audisio granted their wish and ordered Mordini, head of the firing squad, to proceed without Petacci, who would be executed separately.

The prisoners were called to attention and told to *about face*. All except one turned around to face the lake. Decorated World War I veteran Fascist official Francesco Maria Barracu—he had negotiated about the armored car—a black patch over his left eye, stepped forward and pointed to a gold medal in his lapel. As a holder of the Medaglia d'Oro, Italy's highest award for heroism, Barracu insisted he had a right to be shot in the chest instead of the back. Delle Stelle wanted to intervene on behalf of the old soldier, but Audisio shouted to Barracu to turn around and be shot in the back like everyone else. "Those are my orders, and they will be carried out," said Audisio. Barracu obeyed, turned around and, like the others, faced the picturesque Alpine lake surrounded by snowcapped mountains.

When Mordini ordered the firing squad to aim (*Puntate!*), the partisans raised their submachine guns and pointed them at the prisoners. Some of the prisoners gave a Fascist salute. Others said, "Long live Italy!" At the order to fire (*Fuoco!*), the air was filled with the sound of multiple, simultaneous, bursts of automatic gunfire for several seconds. The odor of burning powder and bluish smoke from the submachine guns wafted over the piazza as the prisoners crumpled backwards to the ground. Standard procedure would then have been for Mordini, in charge of the firing squad, to order, "Cease Fire!"; whereupon he, or another, with pistol drawn, would check to assure each prisoner was dead. Any alive would be given the *colpo di grazia*, a pistol shot at point-blank range, usually to the heart.

Satisfied the fifteen were all dead, Mordini, or possibly Audisio, yelled, "Bring out Petacci."

Marcello Petacci had witnessed the executions and was terrified. A rather athletic and powerful man, he threw off the men holding him and ran toward the Dongo Hotel where his "companion" Zita Ritossa and their two sons were staying. He was recaptured and brought toward the railing where the others had been executed. He escaped again, was

Moments before the execution at Dongo.[c]

wounded in the leg, and dove into the lake; but he could not escape the hail of automatic weapons fire. Delle Stelle sent a team out in a boat to retrieve the body.

From her hotel window overlooking the lake, Zita Ritossa watched as her lover, the father of her two sons, was killed. Another onlooker was Aldo Lampredi, co-leader with Audisio of the expedition to Dongo. He viewed the executions and the killing of Marcello Petacci from the town hall balcony overlooking the piazza.

Many bursts of gunfire followed as the partisans fired into the air, celebrating what for them seemed the end of the war.

The covered Fiat truck Audisio had brought to Dongo was loaded with the bodies. Sand was put on the floor of the truck to soak up the blood, and sheets were thrown over the corpses. Audisio had told delle Stelle to have Marcello Petacci buried in a nearby cemetery; but delle Stelle, disgusted by the whole event, disobeyed and had Petacci's body thrown on the truck with the others. Members of the firing squad rode in the same truck with the cadavers when the vehicle with its macabre cargo left the Dongo piazza for Milan after 7 p.m.

Marcello Petacci.[c]

Delle Stelle described the scene in the piazza: pools of blood where the executions had occurred, with the faint smell of gunpowder still in the air.

Of the over fifty Italians arrested at Musso on April 27, 1945, the ones not executed at Dongo were sent to Como the following day, and perhaps eight or twelve, or so, were executed there over the next couple of days before the Allies stopped the summary executions. Most of the prisoners, like Mussolini's secret daughter Elena Curti, were given prison sentences. However, despite the war being over, the partisans were not done. An estimated nine thousand Fascists died in revenge killings in April and May 1945, alone.

On the way back to Milan, the truck from Dongo stopped in Giulino di Mezzegra/Azzano to pick up the bodies of Mussolini and Clara Petacci, which were added to the sixteen already on the truck.

The American 1st Armored Division, part of Fifth Army's IV Corps—the same corps Dad had been in—had almost completely surrounded and sealed off the northern half of Milan. (The Brazilians were surrounding the southern half). With the Allies poised to enter the city, Audisio was fortunate to have the pass signed by OSS officer Daddario. The truck made it through several partisan and Allied check-points and entered Milan about 10:30 p.m.

42

PIAZZALE LORETO

In their car, on the return trip, Lampredi and Audisio discussed what to do with the bodies, something apparently not decided when they set out on their mission that morning. Audisio agreed with Lampredi that since the fifteen high-level Fascists had been killed in retribution for the execution the previous August in the Piazzale Loreto, the *Piazzale Loreto* would be the best place to off-load the truck.

Not far from their destination, they came across a partisan roadblock at a Pirelli tire factory. The partisans there were not Communists, however, but Christian Democrats. Audisio went inside the factory to phone in and request fresh troops, since the men with him were tired. When he went outside, a Christian Democrat partisan captain pointed a pistol at him, disarmed him and put him against a wall. He went through Audisio's pockets, finding a list of the Fascists arrested at Musso. Audisio was accused of being a Fascist posing as a partisan. Meanwhile, Lampredi, Mordini, and one of the Como Communists had taken the car to search for fresh Communist partisans to replace the men in the truck. When they could not find a Communist post nearby, they returned to the truck to find Audisio and his men missing. They themselves were also arrested and disarmed by the Christian Democrat partisans. Lampredi later admitted he feared for his life. Phone calls were made and returned. A senior officer appeared. After about three hours, they were all freed. Audisio, however, irate, disarmed the captain and two other officers, announcing they were under arrest, and put them on the truck.

The bodies were dumped near an *Esso* filling station under construction in the Piazzale Loreto by about 3:00 a.m., and the co-leaders of the expedition proceeded to the partisan military headquarters. Audisio demanded punishment for the partisans who had arrested him. Lampredi, by phone, attempted to contact Luigi Longo, the man who had sent Audisio and him on their mission the previous morning. Longo was finally located at a plant where *L'Unità* and other anti-Fascist newspapers were printed. He was disappointed to learn the bodies had been dumped in the Piazzale Loreto, saying it would profane the place that had become such a symbol—and detract from the memory of the fifteen partisans martyred there the previous August. Lampredi tried his best to justify the decision. Longo asked for a list of the fifteen Fascists executed at Dongo; but after Lampredi had given information on only a dozen, Longo said he had to go—the newspapers were facing a deadline for a special edition, and he needed to write up what he had.

That Sunday morning, April 29, 1945, the headline of the Communist newspaper *L'Unità* announced "patriots" had executed Mussolini and his "acolytes," and their bodies were in the Piazzale Loreto. Those who turned on the radio heard the news that way.

What happened over the next several hours can only be compared to mob scenes in Paris during the French Revolution in the *Place de la Concorde* (called *Place de la Révolution* at the time).

Guards were unable to manage the throngs of people, some in the crowd brandishing submachine guns and pistols. Mussolini had been in power since 1922. A whole generation had grown up under the yoke of Fascism; and many anti-Fascists took revenge on the bodies of the dead Fascists, especially Mussolini. He was kicked in the head so many times his face looked hardly human. Some in the crowd reportedly urinated on the bodies. Others spat. A woman pulled out a revolver and shot Mussolini's corpse five times—once for each of her sons lost under Fascism. Firemen were called and tried to control the crowd with fire hoses.

For Milan, this seemed the time to celebrate the end of the war. Mussolini was dead. Fascism was dead. The Black Brigades were gone. The Germans were defeated. No more secret police. No more Allied bombing. No more Fascist-controlled newspapers and radio. No more rationing and starving. Occasional celebratory bursts of gunfire came from the crowd, as everyone tried to see the bodies on the ground.

One of the founders of Fascism, Achille Starace, was on a morning jog in Milan when he was arrested by partisans and brought to the Piazzale Loreto. He was made to face the rear wall to the right of the filling station building and was executed by a partisan firing squad at about 11:30 a.m.

In order to satisfy the crowd, the bodies of Mussolini and Petacci were hung upside down from the gas station's canopy. Petacci's skirt was tied for modesty. Starace's bloody shirt was removed and his body also hung upside down from the gas station canopy with Mussolini's, Petacci's, and Party Secretary Pavolini's.

As Mussolini hung upside down, his jacket obscured his famous face. The jacket was removed, leaving him in his undershirt; but while his body had been on the ground, his head had been kicked so much the face had become ghoulishly distorted.

Part of the crowd at Piazzale Loreto on April 29, 1945.[c]

Achille Starace before his execution.[c]

Mussolini, Petacci, Pavolini and Starace.[c]

Eventually, three or four more corpses were hung upside down from the gas station canopy next to the four already there. As each one was suspended, the crowd roared with approval. Though some Italians considered the bloody events in Piazzale Loreto barbarous and shameful, most celebrated their relief and joy that a painful era in history had come to an end—the Germans had departed, war with the Allies, at least in

Advance Fifth Army troops arrived during the day.[6]

Milan, was finally over, the Italian civil war was over. In general, the Italian people were proud Mussolini and his regime had paid for their crimes at the hands of their fellow Italians.

Some American soldiers were in the crowd. Troop B of the 91st Cavalry Reconnaissance Squadron, Mechanized, attached to Fifth Army's 1st Armored Division, entered the city on April 29 to verify reports Milan was under partisan control. Signal Corps members photographed and filmed some of the events in Piazzale Loreto, as did newsreel cameramen and photojournalists who worked for newspapers and wire services. Mussolini's death and the desecration of his body were reported by the British Broadcasting Corporation (BBC), which was monitored in the Führerbunker in Berlin. Adolph Hitler, when he heard the news, said the same would not happen to him. He had already dictated instructions his body was to be burned after his death.

The following day, he committed suicide.

Along with the first American troops to enter Milan was a civil affairs officer, Colonel Charles Poletti, a New York lawyer and politician who had actually served for a month as interim governor of the state. He visited Piazzale Loreto, among other sites, during the day and was officially received at the partisan command headquarters that evening, as plans were made for the transition of power from the CLNAI partisans

Photo Dad brought home of the Piazzale Loreto filling station in Milan where Mussolini, Petacci and the others were displayed on April 29, 1945. (The arrow pointing to "Mussolini" is Dad's).[c]

to the Allied Military Government. Of Mussolini's death, he said, "The Italians did what they had to do." The following day, Monday, April 30, 1945, the 1st Armored Division and other elements of Fifth Army's IV Corps, under Major General Willis D. Crittenberger, entered Milan and officially "liberated" the city.

. . .

After the war ended, Dad, at some point in 1945 during his stay in Italy, visited Milan. The Piazzale Loreto had become a prime attraction for the GIs. He brought home a photo of the Esso filling station where Mussolini and the others were hung upside down. I do not know how he acquired it, but he was very, very proud of it.

The photo Dad brought home is unusual. It lists eight names, apparently written from overhead while the bodies were still suspended. However, I have not seen a photo so far that has more than seven bodies. Two of the names were mistakes. One was quite apparent. *Attilio Teruzzi*—the "z's" were written backwards on the filling station canopy—was not executed at Dongo, and in fact survived the war. He was well known for his distinctive beard. The person hung by his heels under Teruzzi's name was likely *Nicola Bombacci*, who was executed at Dongo, and had a beard.

The biggest mystery, however, was the body immediately to the left of Mussolini's under the name *Gelormini*. Sources agree Colonel Giuseppe Gelormini was commander of the National Guard in Milan, but he was not executed at Dongo.

Everywhere I looked, no one seemed to know how his body got to be hung by the feet in Piazzale Loreto on April 29, 1945.

After much research in English and Italian sources, I found an obscure diary in Italian written by a Communist partisan named *Lamberto Caenazzo* that explained what apparently happened. Caenazzo was involved in arresting Gelormini, a rather fat man, on April 28. However, Gelormini protested he had previously surrendered his unit to the partisans in exchange for his personal safety and that of his family. The surrender had been to *Liberal* partisans, not to Communists, so he was released to the custody of two Liberal partisans who vouched for him.

When, like so many thousands of others, Caenazzo visited the Piazzale Loreto the following day, he was shocked to see a body hanging from the filling station canopy next to Mussolini's under the name "Gelormini." He knew immediately the fit body was *not* that of Gelormini, an obese man.

He stated in his diary he did not know at the time whose body was under Gelormini's name but learned later—he did not say how—it was actually *Marcello Petacci* (executed at Dongo but not on the list of fifteen).

So, Mussolini by Caenazzo's account, was hung upside down *between the two Petacci's*, brother and sister, hated by both the Fascists and anti-Fascists alike.

According to Caenazzo, Gelormini, who survived the war, ran for the Italian senate from Genoa in 1948.

[I must say I have my doubts the body under Gelormini's name was Marcello Petacci's, since the body was wearing boots and jodhpur military pants; and Petacci reportedly was shoeless when shot and was posing as a Spanish civilian.]

43

AFTERMATH

It is not clear whether the arrival of the American troops and the new Allied governor, Colonel Charles Poletti, had anything to do with the bodies of Mussolini, and the others being taken down about 2 p.m. on April 29. Some claim Cardinal Schuster ordered it.

A much-photographed, highly controversial autopsy on Mussolini's body was conducted the following morning in Milan by Professor Caio Mario Cattabeni and two associates. At one place, the autopsy report said Mussolini was hit by *seven* bullets while alive, at another *nine*. All bullets supposedly had both entrance and exit wounds, but no information was given on the caliber of the bullets. In addition, Mussolini had postmortem bullet wounds. The state of the body supposedly was more consistent with death *earlier* than 4:10 p.m. on April 28, as had been reported. No autopsy was done on Clara Petacci. Both were supposedly buried anonymously in a Milan cemetery, Mussolini in a potter's grave and Petacci under a false name. Mussolini's body was stolen a year later by a small group of Fascists, recovered, reburied again and finally returned to his widow Rachele in 1957. She had him buried in a family tomb in Predappio, where he had been born.

Clara Petacci's body was exhumed in 1947. A Milan newspaper reported eleven years later that the body had been examined at that time, and two 9mm slugs were supposedly found, whereas she supposedly had been shot by the same 7.65 mm submachine gun that killed Mussolini. As I indicated previously, conspiracy theories continue to this day as to how and even where Mussolini and Petacci were killed.

44

THE DONGO TREASURE

On April 29, 1945, while the shocking events of Piazzale Loreto were going on in Milan, Dad's battery was still stuck in Malcesine on the east shore of Lake Garda due to the damaged road and tunnels north of there. Enemy activity was observable on roads on the opposite shore departing from the Salò/Gargagno area that had been the seat of Mussolini's government, resulting in fire missions for Dad's battery. Orders were given for a ten-man raid from the 10th Mountain Division to cross the lake in broad daylight in a motorboat and capture what had been Mussolini's residence, the Villa Feltrinelli in Gargagno. The Allied troops around Lake Garda apparently were unaware Mussolini and his family had departed the area for Milan and Lake Como a number of days earlier. When Colonel Darby, former commander of Darby's Rangers and now assistant division commander, heard of the planned operation, he did not like it.

Colonel Darby changed the plan to require an entire company, plus other elements, to accomplish the mission. Instead of a daylight raid, the new plan called for a predawn attack the following morning, April 30; and instead of a motorboat, the larger attacking force would use large

DUKW amphibious craft to cross the lake and would land a mile or two north of the villa. While the mission depended in part on silence, it is likely Dad's battery, across the lake, was *on call* in the event the raiders ran into resistance and needed counter-battery firepower—though the villa itself was probably out of range.

The mission planned by Colonel Darby went off without a hitch, however. An old Italian man on a bicycle approached the Americans and told them the Germans had all departed, which proved to be true. The Villa Feltrinelli had been abandoned. The soldiers from Company K of the 85th Mountain Infantry found the table set, clothes in the closets—everything much as it must have been when Mussolini, his wife Rachele, and two youngest children lived there.

One room was full of medications, like a pharmacy. The inboard boat at the pier and the limousines in the garage had been sabotaged, so as not to be usable. Nevertheless, the soldiers took advantage of the beds and the kitchen to make a comfortable bivouac.

Mussolini left behind many uniforms, awards, decorations, and souvenirs. He had received swords from Hitler and Emperor Hirohito of Japan. He played violin and had a Stradivarius. Some items were taken

Villa Feltrinelli on Lake Garda as it looks today. Now a hotel.

by the soldiers as "spoils of war" and were mailed back home. Supposedly, no great riches were found, but already the fall of Mussolini's Italian Social Republic was beginning to result in a question—*where is the money?*

In the days that followed, the Allied Military Government began looking for the riches of Mussolini's regime, the large amount of Italian and foreign currency, gold, jewelry, and other valuables Mussolini and the hierarchy of his government had with them when captured at Musso and Dongo on April 27, 1945, generally called the *Dongo treasure* or the *Dongo gold*. It all seemed to have disappeared in the days following the death of Mussolini. The money supposedly on the pickup truck that left Milan headed for Como when Mussolini did on April 25, was never seen again. The items taken from Mussolini by Urbano Lazzaro of the Puecher detachment of the 52nd Garibaldi Brigade in Dongo and deposited in a bank in Domaso, north of Dongo, disappeared, as did the 33 million lire given by German officers to an associate of Lazzaro's and deposited in the same bank. Wedding rings and gold jewelry donated to the Fascist government by patriotic Italian women at the time of the Ethiopian war, thrown into a river by the fleeing Germans and retrieved by two fishermen, also disappeared.

The Fascists captured at Musso were ready to leave the country and brought with them untold French and Swiss francs, Spanish pesetas, bank notes, jewelry, and other valuables. The Allied Military Government estimated the missing treasure to be worth over $66 million.

Several members of the 52nd Garibaldi Brigade were murdered, including a man and a woman involved in taking Mussolini and Clara Petacci to the De Maria house in the early morning hours of April 28. Giuseppe Frangi, one of the men who guarded Mussolini and Petacci at the De Maria house and guarded the bodies later, was soon killed. Michele Moretti, whose gun was used to kill Mussolini and Petacci, was at one time accused of theft and fled the country. Inquiries lasted a dozen years with no clear conclusion.

Most authorities believe the bulk of the Dongo Treasure ended up in the hands of the Italian Communist Party, which spread its influence and built a new party headquarters in Rome after the War.

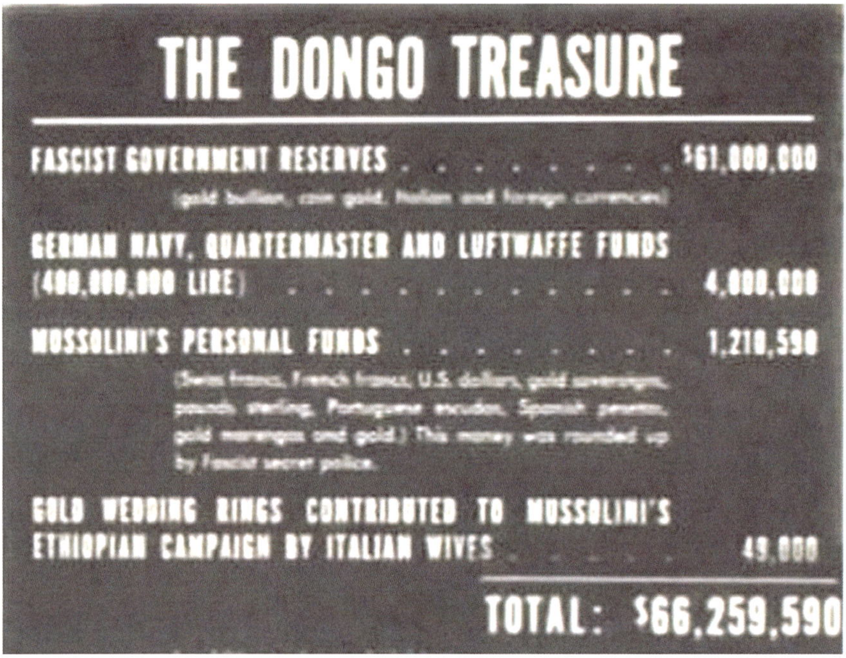

THE DONGO TREASURE

FASCIST GOVERNMENT RESERVES $61,000,000
(gold bullion, coin gold, Italian and foreign currencies)

GERMAN NAVY, QUARTERMASTER AND LUFTWAFFE FUNDS
(400,000,000 LIRE) 4,000,000

MUSSOLINI'S PERSONAL FUNDS 1,210,590
(Swiss francs, French francs, U.S. dollars, gold sovereigns, pounds sterling, Portuguese escudos, Spanish pesetas, gold marengos and gold.) This money was rounded up by Fascist secret police.

GOLD WEDDING RINGS CONTRIBUTED TO MUSSOLINI'S
ETHIOPIAN CAMPAIGN BY ITALIAN WIVES 49,000

TOTAL: $66,259,590

Allied Military Government estimate of missing treasure.

One account claimed the gold was almost immediately melted down in one of the furnaces of the Falck steelworks in Dongo for the Communists.

Among the items missing from the bank in Domaso, part of the contents of the large leather briefcase Mussolini had with him when captured, was a file containing correspondence from British Prime Minister Winston Churchill.

When Britain was fighting the Germans alone, before Italy declared war on Britain in June, 1940, Churchill supposedly offered Mussolini some embarrassing concessions in East Africa to keep Italy neutral. Mussolini had also acted as an intermediary in Germany's negotiations with Britain in 1940 over possible surrender. Most Americans are unaware of the fighting that was carried on for a year and a half between Britain and Italy early in the War over Italian East Africa, to guarantee Britain's access to the Red Sea, and thence Suez.

Mussolini's case also contained correspondence from Adolph Hitler. Mussolini apparently intended to use the documents to negotiate with the Allies for his freedom. The documents also supposedly contained an incriminating report of "acts of sexual deviance" (i.e., homosexuality) by Crown Prince Umberto, heir-apparent to the monarchy. While some of the documents Mussolini had with him have seen the light of day, most have not. Some historians have speculated the Communist Party kept them, others think the Allies prevented them from being published. Some believe Mussolini had documents dumped into Lake Garda before he left for Milan. In short, the disappearance of the Dongo treasure and secret documents gave conspiracy theorists enough fodder to last several generations.

45

OPERATION SUNRISE

While the bodies of Mussolini, Clara Petacci, and the others were hanging upside down from the canopy of the Esso station in Piazzale Loreto in Milan, an event with an even bigger impact on Dad and the other Allied soldiers than the end of the Fascist regime was occurring in Southern Italy.

I previously mentioned how, in March, Waffen SS General Karl Wolff, head of the SS in Italy and Western Austria, had tried to negotiate a surrender with Allen Dulles, head of the American Office of Strategic Services in Switzerland, but was unable to obtain the cooperation of the head of the German army in Italy, General Heinrich von Vietinghoff. When the Soviets learned of the peace talks, they feared the Americans and British would allow German soldiers from Italy to join the fight against them on the Eastern Front. Stalin complained to the President, and Dulles was ordered to break off the talks, known as *Operation Sunrise*.

Operation Sunrise was really what might be called a *back-channel* attempt at ending the war in Italy. It began with an international businessman

Max Husmann, Max Waibel, and Baron Luigi Parrilli. Three unlikely friends who helped end a war.[c]

from Genoa named Baron Luigi Parrilli, who, before the war, had been the Italian representative of the Kenosha, Wisconsin-based auto manufacturer *Nash Motors Company*, which merged in 1937 with Detroit appliance manufacturer *Kelvinator* to become *Nash-Kelvinator Corporation*. Beginning with a chance meeting in a hotel breakfast room in Genoa, Parrilli befriended an SS lieutenant named Guido Zimmer, and learned the Germans planned to destroy much of the infrastructure of Northern Italy, including the port of Genoa, before they departed. (The Allies were well aware of this threat. A primary mission of the partisan militias was to preserve Italian ports, factories, power plants, and other infrastructure vital to the economy, as the Allies advanced. For example, before coming out of the hills, delle Stelle's 52nd Garibaldi Brigade was to make sure a power plant north of Lake Como was not destroyed by the departing Germans).

Through Lieutenant Zimmer, subsequently promoted to captain and major and transferred to Milan, Parrilli made contact with the colorful Eugen Dollmann, an urbane translator and assistant to SS General Karl Wolff. Dollman had translated for Mussolini and Hitler, and seemed to know everyone—he had even previously met Parrilli years earlier at a social event. General Wolff, as it turned out, was looking for some way to communicate with the Allies.

Parrilli, at great risk—the partisans thought he was a Fascist and the Fascists suspected him of working for the Allies—decided to do what he could to bring the Germans and the Allies together for the sake of Italy. He contacted an old acquaintance of his in Switzerland named Max Husmann, a Ukrainian Jew who ran a celebrated international boarding school called the *Institut Montana* (still in existence) outside Zug, near Lucerne. Parrilli, who spoke German and French besides Italian, had a relative who had attended Husmann's school. Husmann was friends with a Swiss intelligence officer in Lucerne named Max Waibel. Waibel put Parrilli and eventually General Wolff himself in contact with the OSS chief in Switzerland, Allen Dulles, and with Dulles's top agent, German-born Dr. Gero von Schultze Gaevernitz, another person Dollmann had met years before.

The Swiss had an interest in seeing the war end. The landlocked country had two primary commercial routes to the sea. The Rhine River flowed north from the Basel area through Germany and Holland to ports on the North Sea, like Rotterdam. Access to the Mediterranean, to the south, was by rail to the Italian port of Genoa. The fighting in Europe in spring of 1945 put both in peril. In addition to its commercial concerns, the neutral Alpine nation of about four million had also become a "lifeboat" for thousands of war refugees who could not go home until the fighting ended.

The Americans at first did not put much stock in the peace initiative from an Italian businessman and a Swiss headmaster; but when General Wolff, in a show of faith at Dulles's request, arranged to have two major partisan leaders—including future Prime Minister Ferruccio Parri—released, the negotiations were taken seriously.

On April 21, 1945, after two months of negotiations, including several

Americans Dr. Gero von Schultze Gaevernitz and Allen Dulles, Bern, Switzerland station chief of the Office of Strategic Services (forerunner of CIA), in 1945.c

meetings in Switzerland with General Wolff himself, Dulles received orders from the Joint Chiefs of Staff to break off contact. President Truman had succumbed to the Russians and the British, who supported the Russian position on Operation Sunrise. Dulles and the higher-ups in the Mediterranean Theater of Operations were upset. Prolonging the war in Italy meant not only more death and destruction but also an opportunity for the Russians from Hungary or Communists from Yugoslavia to take the important Italian Adriatic port of Trieste and possibly more of Northern Italy with the help of Communist partisans.

Stalin, of course, was paranoid, suspecting the English and Americans might do what Russia itself had done in World War I.

After the October 1917 revolution in Russia, Lenin made a separate peace with Germany, which allowed the Germans to transfer fifty divisions from the Eastern Front to France. A spring 1918 offensive, with American troop strength still building, came close to winning that war for Germany.

Stalin also was aware Hitler hoped the British and Americans would view the Red Army advances as a threat to post-war Europe and would turn against the Soviets. So, Stalin insisted a Soviet official be present for any peace negotiations undertaken by the Americans and British with the Germans.

Ironically, the same day Dulles was told to end Operation Sunrise, he learned from Baron Parrilli that General Heinrich von Vietinghoff, head of the German Armed Forces in Italy—after his headquarters at the spa town of Recoaro Terme, seventy miles east of Lake Garda, had been heavily bombed on April 20—was finally persuaded by SS General Karl Wolff to surrender. General Wolff and two adjutants, one from the *Waffen SS* and one from the *Wehrmacht*, would be arriving in Switzerland in two days to travel to Allied headquarters for the surrender.

Having moved from Algiers in July of 1944, while Dad was still on Corsica, Allied headquarters for the Mediterranean Theater—originally called the "North African Theater"—was at that time in Southern Italy, at the king's summer palace in Caserta (said to be larger than the *Versailles* palace), about thirty miles outside of Naples. It was here that British Field Marshal Harold Alexander, Supreme Allied Commander in the Mediterranean Theater, had his headquarters and where a Russian delegation worked with the British and Americans regarding all fronts in the Theater, including Italy.

In the previous negotiations, the Germans had understood the actual surrender would have to occur at Caserta with a Russian representative present. Saying he would undertake no further contacts, Dulles asked Caserta for *reconsideration* of the decision to terminate Operation Sunrise. He also emphasized the critical role of the Swiss in the negotiations, and how disappointed they would be if the decision to terminate Sunrise were not reversed. When General Wolff and the two adjutants, dressed in civilian clothes, arrived in Switzerland, they were put up at the villa of Swiss Intelligence officer Max Waibel near Lucerne to await word from Dulles—who, of course, could not speak to them, except to leave word that they sit tight. Field Marshal Sir Harold Alexander, Supreme Allied Commander in the Mediterranean, recommended to his superiors in London and Washington the restriction be lifted, so arrangements could be made for the Germans to come to Caserta and surrender.

Days went by.

Palace at Caserta, outside Naples, Allied headquarters for the Mediterranean Theater.[c]

Frustrated, General Wolff departed Switzerland late on April 25, leaving behind the two junior officers at Max Waibel's villa near Lucerne with full authority from himself and General von Vietinghoff to surrender.

As I previously related, an SS villa in Cernobbio on Lake Como, where General Wolff stopped, was surrounded by partisans on April 26. The partisans had neglected to cut the phone line, however, and a Swiss intelligence agent was able to reach the SS general. Wolff was rescued by a hastily-organized OSS team led by an agent named Donald Jones, well known to the Italian partisans as *Scotti,* and taken back to Switzerland in the early morning hours of April 27. From there, very grateful to the Americans and the Swiss, he eventually was delivered to the Austrian border, and headed through Austria for his new headquarters in Bolzano in the Italian Alps, where von Vietinghoff also had moved his headquarters. Wolff's previous headquarters was on Lake Garda, not far from Salò, but the recent advances of the 10th Mountain Division had made staying there impossible.

The move of Wolff's headquarters prompted a change in radio contact. On April 13, Dulles and Wolff had created a means of direct contact to avoid the problems messengers like Baron Parrilli and others were having at the Swiss border and traveling through partisan areas. A shortwave radio had been secretly installed at the office of Major Guido Zimmer in Milan for contact with Dulles in Switzerland. Zimmer was the friend of Parrilli's who worked for General Wolff. The radio was operated by a courageous twenty-six-year-old Czech, a former law student and Dachau concentration camp escapee who spoke nearly flawless German and had been an informant for the Czech underground before going to work for the American OSS. His name was Vaclav Hradecky, but the OSS called him "Little Wally."

As the Germans left Milan he went back to Switzerland, but arrangements were soon made to have "Little Wally" and his radio transported from Switzerland to Wolff's new headquarters in the Palazzo Ducale in Bolzano for the purpose of direct contact with Caserta—in case the surrender could still be arranged.

In the Operation Sunrise negotiations in Switzerland, the Germans had agreed not only to lay down their arms, and become prisoners of war, so as not to participate on the other fronts, but also to refrain from acts of sabotage against Italian infrastructure. Emphasis was on trying to keep order. The Germans were to be considered prisoners of war and entitled to the protections of the Geneva Convention. The Allies would assume responsibility for protecting the Germans from the partisans, who now seemed a constant threat to the Germans.

The morning of April 27, 1945, Dulles finally received a triple-priority cable lifting the restrictions on Operation Sunrise and advising him to arrange for the two German officers to fly to Caserta via an Allied air-base at Annecy, France, not far from Geneva.

The two officers, in plain clothes, were Lieutenant Colonel Viktor von Schweinitz, an aide to General von Vietinghoff of Army Group C, on behalf of the Armed Forces, the *Wehrmacht*; and Major Eugen "Max" Wenner, an aide to General Wolff, on behalf of the *Waffen SS*. On April 28, having crossed the French border at Annemasse near Geneva, they boarded Field Marshal Harold Alexander's C-47 (military version of the DC-3) at the Allied air base at Annecy and flew in stormy weather to Caserta. Although they were only field-grade officers, below the rank of general, the two Germans had the full authority of their superiors and therefore were referred to as *plenipotentiaries*. They were accompanied by Allen Dulles's agent Gero von S. Gaevernitz.

Upon their arrival at Caserta late in the afternoon, they were met by American Major General Lyman L. Lemnitzer and English Major General Terence S. Airey, both of whom had previously participated in Operation Sunrise negotiations in Switzerland in plain clothes and under assumed names.

Staged photo of "Little Wally" taken at Gen. Wolff's headquarters in Bolzano after the war. The SS officer with him was Maj. Guido Zimmer, friend of Baron Parrilli's, who worked for SS Gen. Wolff. While in Milan and Bolzano, "Little Wally," a Czech who was a Dachau escapee, wore a Nazi uniform and pretended to be a German soldier.

The Germans were taken to a separate bungalow on the palace grounds where informal discussions took place over tea. Three hours later, the plenipotentiaries attended a conference in the palace with a number of Allied officers, including a Russian major general as an observer. There, Lieutenant General Sir William D. Morgan, chief of staff to Field Marshal Alexander, gave them three copies of the proposed surrender document and directed them to study it in their quarters and return for questions and discussion at 9 p.m.

At that meeting, the German representatives demanded all surrendering officers and military police be allowed to retain side arms so as to keep order, and General Morgan agreed. While the surrendering Germans were to abandon their equipment, General Morgan also agreed to allow them to use their military trucks to carry rations to feed their troops. But he denied a request for instant demobilization, as had been done at the end of World War I, insisting the surrendering German troops would have the status of prisoners of war. Afterwards, von Schweinitz, Wenner, Gaevernitz and

Caserta surrender. At left in plain clothes Lt. Col. Viktor von Schweinitz representing Gen. von Vietinghoff of Army Group C, German armed forces (*Wehrmacht*) and Maj. Eugen Wenner representing Gen. Karl Wolff of the Waffen SS. On the right, Gen. William D. Morgan representing Field Marshal Harold Alexander, Supreme Allied Commander in the Mediterranean, with American Gen. Joseph McNarney and various other Allied representatives, including a Russian (Maj. Gen. Aleksei Kislenko). With the handkerchief in his lapel pocket is OSS agent Gero von S. Gaevernitz.[c]

General Lemnitzer stayed up most of the night going over the surrender document point by point. An effort was made to transmit the terms through Little Wally in Bolzano for review by General von Vietinghoff, but weather conditions were such that the radio signal was garbled.

In the afternoon of Sunday, April 29, 1945, shortly after the bodies of Mussolini, Petacci and the others were taken down in the Piazzale Loreto in Milan, a signing ceremony took place in the palace at Caserta to end the war in Italy.

The parties signed five copies of the surrender document, in English and German, three of which were given to von Schweinitz and Wenner, who spoke both languages. Interestingly, Lieutenant Colonel von Schweinitz's grandmother had been an American and was actually a descendant of John Jay, author of five of the *Federalist Papers* and first Chief Justice of the United States Supreme Court.

The surrender document included orders in separate appendices for German land, naval, and air forces. When General Wolff was at the villa in Cernobbio on April 26 and was surrounded by partisans, awaiting rescue by the OSS, he obtained from Italian defense minister General Rodolfo Graziani, also at the villa at the time, a *proxy* to surrender all forces of the

Italian Social Republic. Thus, Major Wenner's authority, through General Wolff, also included surrender of all the Italian Fascist armed forces.

Under terms of the surrender, all hostilities were to cease at *noon* Greenwich Mean Time (2 p.m. in Italy) on May 2, 1945. This was to allow the German plenipotentiaries time to get back to their headquarters in Bolzano with the surrender document and for the Germans to get word to their troops to stop fighting as of the time of the cease-fire. General Morgan briefed members of the British and American press afterwards and asked them to keep the surrender a secret for three days, *which they did*. The Allies were apprehensive the premature leak of surrender would cause their own troops to let up the military pressure, possibly allowing more German troops to flee from Italy to fight on the other fronts.

Shortly after the signing ceremony at Caserta on April 29, von Schweinitz and Wenner, along with OSS officer Gaevernitz, having had little opportunity to sleep, flew back to France. Fortunately, Caserta succeeded in radioing the terms of the surrender to Little Wally in Bolzano for Generals Wolff and von Vietinghoff.

After their nearly four-hour flight, von Schweinitz, Wenner and Gaevernitz landed in France and crossed the Swiss border, but missed a train from Geneva to Bern, and finally were driven by car to Allen Dulles's apartment in Bern, where they arrived at about midnight.

Within an hour, they were on their way, during the early morning hours, in an American car headed for the Austrian border. There a car sent by General Wolff awaited the two Germans to take them to Bolzano. However, Wolff also sent a warning of a plot by Franz Hofer, Nazi head of government in Austria, to arrest them once the car reached Innsbruck, Austria enroute to the Brenner Pass. Hofer wanted to surrender Austria on his own terms and was now trying to sabotage the Caserta surrender. To avoid Innsbruck, von Schweinitz and Wenner were taken on a less-direct, more mountainous, route back to Italy through the Resia/Reschen Pass (where Dad would soon be stationed).

At that point Bolzano was full of drama, and the surrender far from a sure thing.

46

APRIL 30, 1945

While von Schweinitz and Wenner were driving on rural Alpine roads in Austria and Northern Italy, to Bolzano, plenty of fighting was still going on around Lake Garda, where Dad's battery was stuck at Malcesine on the eastern shore, unable to move north due to damage to the shoreline road and tunnels.

I already mentioned how a company from the 85th Mountain Regiment easily captured Mussolini's villa on the southwest shore of Lake Garda in the early morning hours of April 30. At the north end of the lake, however, in the towns of Riva del Garda and Nago-Torbole, the Germans mounted a fierce rear-guard action—even bringing in Tiger tanks—to facilitate their retreat north toward Trent and Bolzano, the German headquarters in Italy. (10th Mountain Division commander General George P. Hays was already making plans for attacking Bolzano). Dad's battery fired about a hundred rounds in support of the doggedly advancing 10th Mountain Division troops on April 30.

In Berlin, on the afternoon of April 30, 1945, with Russian forces closing in, Adolf Hitler simultaneously bit down on a cyanide capsule and shot himself in the head in his underground bunker. He left orders that his body and that of Eva Braun, whom he had married the day before and who also committed suicide (by cyanide), be brought up outside of the bunker, doused with gasoline, and burned beyond recognition. (Dental records were all that identified his remains). Thus, three wartime leaders—Roosevelt, Mussolini, and Hitler—all died in April of 1945.

In his final days, Hitler turned on his closest associates, such as his ambitious heir-apparent Hermann Göring and SS chief Heinrich Himmler. He wanted no one to have the power he had as Führer. In his will, dictated to his secretary the day before he died, Hitler named his propaganda minister Joseph Goebbels chancellor, or head of government, and Admiral Karl Dönitz as president of the Reich. However, the day after Hitler's suicide, when the Russians refused Goebbels' offer of a cease-fire, the new chancellor and his fanatical wife Magda poisoned their six children with cyanide and committed suicide themselves. This left Admiral Dönitz as head of government after May 1, 1945.

The death of Hitler, not announced until late in the evening on May 1, was expected to make it easier for the Allies, because it meant the hard-line Nazis no longer had to abide by their *oath* to fight for Hitler to the death. However, the power vacuum created by Hitler's death, and orders he made in anticipation of his death, caused a great deal of chaos in the upper ranks, raising questions as to whether the recently concluded surrender in Italy would actually come to pass.

. . .

Just before 6 p.m. on April 30, a tragedy occurred on Lake Garda.

As fighting died down in Torbole, at the northeast end of the lake, early in the afternoon, the 10th Mountain Division's 86th Mountain Infantry Regiment set up a command post right on the lakeshore. The celebrated new 10th Mountain Division assistant division commander, Colonel William O. Darby, was on a stone walkway outside the new command post, talking to the 86th's commander, another officer, and a sergeant-major, when a single German harassing-fire artillery round, probably from an 88, burst overhead, killing the sergeant-major and wounding Darby and the other two officers. Colonel Darby was taken to an aid station inside the command post, where he died of his wounds

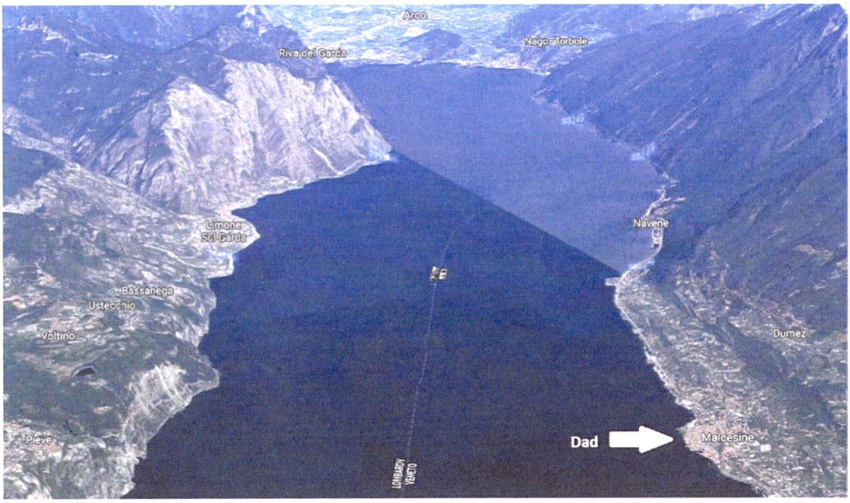

Google Earth shadowy view of Northern end of Lake Garda, looking north. Dad's battery could not progress north of Malcesine due to damage to the road and a series of tunnels. The north end of the lake was up to 1,000 feet deep. (©Google 2017).

in less than an hour. The fluke death of this brave soldier, who had been such an inspiration to those who served under him, was a shock to all his superiors, and I am sure to Dad, whose battalion had been a part of Task Force *Darby*.

In an unprecedented move, thirty-four-year-old Colonel Darby was later honored by being posthumously promoted to *Brigadier General.*

More tragedy was to come.

With the northern end of its over thirty-mile length in the Alps and its southern end closer to the Po Valley, Lake Garda was notorious for having its own wind system. Orders were given for 10th Mountain Division's Battery B of the 605th Field Artillery Battalion (Pack) that night to cross the lake to the west side with their 75mm pack artillery in three DUKW amphibious craft. The first two DUKW's left without a problem, but the operator of the third said the vessel was overloaded, considering the artillery piece, ammunition, a heavy machine gun, and the twenty-five personnel on board with all their gear. Despite his objections, the driver was ordered to proceed.

With the vessel very low in the water, a wind came up and waves came over the sides. In panic, the men, none wearing life vests, started jettisoning their packs and the ammunition for the howitzer—but it was too late. The DUKW swamped and sank, and twenty-four men drowned in the chilly, extremely deep Alpine lake. The only survivor, PFC Thomas Hough, a former lifeguard, managed to stay afloat for forty-five minutes before he was picked up by two infantrymen in a small boat. The others were among the over seventy thousand American service men and women killed in World War II whose remains were never recovered or never identified.

As nearly as I can tell, Dad's battery did not have any more fire missions after April 30. With the entire lake now in Allied hands, the 126th Mountain Engineers could work aggressively on rebuilding the road and clearing and repairing the tunnels so Dad's battery could move again. The 10th Mountain Division proceeded toward Arco, the next major town in the valley north of Lake Garda, on the way toward Trent and Bolzano.

Hitler feared a plan of retreat would damage morale, and so Army Group C was to keep fighting. The only retreat allowed would be to a stronger defensive position. In fact, the time for any strategic withdrawal of German forces had passed. Allied air power had so damaged the roads and railroads to the Brenner Pass, the key escape route out of Italy, that German Army Group C was effectively trapped in Northern Italy.

At Bolzano the constant Allied military pressure on Army Group C was contributing to what could only be described as *chaos*.

47

CRISIS IN BOLZANO

On April 28, 1945, while Lieutenant Colonel Viktor von Schweinitz was flying toward Caserta with authority to surrender on behalf of Army Group C's commander, General Heinrich von Vietinghoff, he was unaware Hitler was about to make Army Group C, previously directly answerable to Berlin, now subject to the command of Field Marshal Albert Kesselring (along with several other Army Groups in the south). Kesselring, who secretly on a previous occasion had sanctioned General Wolff's overtures through the Swiss, was unaware of the plenipotentiaries traveling to Caserta. He had met earlier that day with von Vietinghoff in Innsbruck, Austria, saying he was *opposed* to any surrender that was without his approval. Von Vietinghoff had relied on SS General Wolff being at the meeting, but Wolff had been delayed by his near-capture and rescue at Cernobbio and did not attend. With General Wolff absent, Von Vietinghoff did not tell Kesselring directly of the surrender. Later, he was dismayed and told his chief of staff, General Hans Röttiger, he could no longer go through with the surrender. General Röttiger accused von Vietinghoff of a "lack of personal courage."

Palazzo Ducale of the Prince of Pistoia, SS Gen. Wolff's headquarters in Bolzano, Italy.[c]

The following day, April 29, Franz Hofer, the top Nazi in Austria, who was opposed to any surrender that did not involve him personally, informed Kesselring of the Caserta surrender and accused von Vietinghoff and Röttiger of treason. The next morning, April 30, Kesselring removed von Vietinghoff and Röttiger from their positions. Von Vietinghoff followed orders to report to near Lake Carezza/Karersee, 17 miles outside Bolzano, to await court martial, but Röttiger disobeyed and remained in Bolzano, ostensibly to assist in the transition to von Vietinghoff's replacement, General Karl Schultz, and his own replacement, General Fritz Wentzell. Kesselring was strictly *Wehrmacht* and could not directly punish General Wolff, who was SS, but he reported Wolff for investigation to SS General Ernst Kaltenbrunner, no friend of Wolff's. That night American bombers attacked Bolzano. Some of the bombs fell close to General Wolff's headquarters. Little Wally radioed Dulles who asked Caserta to spare the part of the city where Wolff was located.

Shortly after midnight on May 1, about twenty-four hours after leaving Allen Dulles's apartment in Switzerland, von Schweinitz and Wenner finally arrived in Bolzano with the Caserta surrender documents. The deadline for the cease-fire was fast approaching, and General Röttiger, wanting to go through with the terms of the surrender, decided to arrest the newly arrived Generals Schultz and Wentzell, who were in the process of taking over Army Group C. With Generals Schultz and Wentzell under house arrest and General von Vietinghoff awaiting court martial elsewhere, General Röttiger assumed command of Army Group C and ordered all communication with Germany shut down. He then summoned the commanding generals of the two armies in Army Group C. They both denied Röttiger's authority and refused to attend a meeting called by Röttiger to implement the cease fire.

Röttiger became despondent. At around noon, an officer from Röttiger's staff barged into General Wolff's office to tell him General Röttiger was debating shooting himself. General Wolff raced over to General Röttiger's office and talked him out of suicide. He suggested trying to talk Generals Schultz and Wentzell into going along with the surrender. The two generals were released from house arrest and General Wolff alone talked to them at length about the hopeless military situation in Italy. The two of them finally came around and agreed to the surrender—*but only if Field Marshal Kesselring approved*. At that point, Wolff called in Röttiger, who apologized, and the two generals forgave him for having arrested them and shook his hand.

The commanding generals of the two armies in Army Group C, who were closest to the military situation, having only a fraction of the troops they had three weeks previously, and very little in the way of heavy guns, favored the surrender. They were advised Schultz and Wentzell were now back in command, and communications were restored to Kesselring's headquarters outside Munich, in Germany.

A meeting of the army generals, plus a *Luftwaffe* general and a vice-admiral of the German navy, was set for 6 p.m. at army headquarters in Bolzano to discuss the military situation in Italy and the surrender. All of the military commanders agreed on the hopelessness of the situation, and General Schultz, now back in command of Army Group C, promised to call Field Marshal Kesselring and urge compliance with the surrender, but said he would not implement the surrender without Kesselring's approval. The commanders of the two German armies in

Italy were not prepared to surrender without approval from the head of Army Group C, General Schultz.

It was now the evening of May 1. The surrender signed at Caserta on April 29, 1945 was due to go into effect at 2 p.m. the next day, May 2. News of Hitler's death the previous afternoon had not yet arrived. The cease-fire appeared dead without the consent of Field Marshal Kesselring, who seemed very much opposed. Unfortunately, General Schultz and SS General Wolff were unable to reach him that night.

Meanwhile, Caserta wired Little Wally in Bolzano about 8:00 p.m. The Allies needed time to get word of the cease-fire to their troops in the field. Was the surrender on—*or not*? General Wolff, still trying desperately to get hold of Kesselring, promised an answer within two hours. Kesselring's chief of staff promised to respond to Bolzano by 10:00 p.m., but never got back. Two hours went by, and Kesselring still could not be reached.

Finally, at about 10:30 p.m., General Traugott Herr, commander of the German 10th Army, told a staff officer to give to the troops under his command an order for a cease-fire to go into effect at 2 p.m. the following day. General Joachim Lemelsen reluctantly did the same for 14th Army, as did General Maximilian von Pohl of the *Luftwaffe* and General Wolff for the SS troops under his command.

Despite the actions of the generals of the two armies under him, however, General Schultz was not prepared to go along with the surrender unless Kesselring approved. He called Kesselring's headquarters again, and was told the field marshal was inspecting troops and would not return until midnight. General Wolff sent a message to Caserta via Little Wally that the major units would surrender—but consent had not been obtained from General Schultz of Army Group C, or from Field Marshal Kesselring, who was over General Schultz. At about that time, word came from German radio that Hitler was dead, thus freeing the German officers of their oath to fight to the death for the Führer.

Starting at just before midnight several orders reached Bolzano indicating Kesselring was aware of the surrender plot. First was an order to arrest *Luftwaffe* General von Pohl, which was ignored by an officer on the general's staff who also favored the surrender. Next came orders to arrest von Vietinghoff (who was already awaiting court martial), General

Röttiger, and Lieutenant Colonel von Schweinitz, who had signed for von Vietinghoff at Caserta. General Wolff got the officers away from General Schultz's headquarters and advised they go to their own headquarters and prepare to defend themselves.

General Wolff returned to his office and learned an order had been given to an army tank unit to surround his palazzo. He ordered seven SS police tanks to guard the park around the palazzo and alerted mobile guards, several hundred armed men, to protect him. He also had Little Wally radio Caserta and advise a parachute drop of Allied troops may be necessary.

At about 2:00 a.m. on May 2, before the German army and the SS went to war against each other in Bolzano, Wolff's phone rang. It was Field Marshal Kesselring saying he was aware German troops in Northern Italy were given unauthorized orders to surrender and knew who was behind the treason. He accused Wolff of leading the mutiny, which Wolff did not deny. Wolff appealed to Kesselring to acknowledge the military reality and urged him to approve the surrender. He argued the surrender in Italy would benefit post-war Germany by preventing advances into Italy by the Soviets or the Communist Yugoslavs. Kesselring felt his authority had been questioned. He had wanted time to move some of his troops facing the Russians so that they eventually could surrender to the Americans. However, he admitted the military situation in Italy, as elsewhere, was hopeless. After about two hours, Kesselring eventually agreed to make a decision within a half hour.

At about 4:30 a.m. General Schultz heard from Field Marshal Kesselring. The surrender was *approved* and the orders for the arrests of Generals von Pohl, von Vietinghoff, and Röttiger and Lieutenant Colonel von Schweinitz were rescinded. The cease fire looked like it was still on for that afternoon.

A few hours later a politician arrived in Bolzano: Rudolf Rahn, who had been the German ambassador to Mussolini's government and who really ran the Italian Social Republic. When he was advised of the situation, he wanted to correct a difficulty regarding the surrender. At the time of the surrender in Caserta on April 29, Lieutenant Colonel von Schweinitz signed on behalf of General von Vietinghoff as commander of Army Group C. Ambassador Rahn feared hard-line German commanders

Photo taken in Bolzano about a week and a half after the surrender. From left, Gen. Hans Röttiger, OSS officer Gero von S. Gaevernitz, SS Maj. Eugen "Max" Wenner (in background—signer at Caserta), Gen. Heinrich Von Vietinghoff, SS Col. Eugen Dollmann (in background), and SS Gen. Karl Wolff. (*NARA*).[c]

might argue the removal of von Vietinghoff from command invalidated the surrender. He called Kesselring, who, with reluctance, reinstalled von Vietinghoff as commanding officer of Army Group C.

At 2:00 p.m. local time (noon GMT) on May 2, 1945, the war in Italy officially came to an end, though most Allied soldiers were not aware of it until a few hours later.

48

THE WAR IS OVER

I do not recall Dad telling me how he heard the war in Italy was over. I imagine word traveled fast the late afternoon and early evening of May 2, with shooting in the air, launching of flares, and much celebrating. Dad's battery had been stuck in the vicinity of Malcesine on the east shore of Lake Garda since April 28 due to the Germans having blown several tunnels and having damaged the road north of there. Attached to the 10th Mountain Division and assigned to Task Force Darby, Dad's battery supported operations of the division within the seven-mile range of the howitzers. However, Dad's battery did not have any fire missions after April 30 as the division started moving operations north of the lake. In the meantime, the 126th Engineering Battalion was doing its best, with bulldozers and other equipment, to clear the blocked tunnels and rebuild the road, so Dad's battery could get moving again.

What I do remember Dad telling me about the end of the war was something he had not expected. A lot of the soldiers, when they heard the news, started crying.

"Tears of joy?" I suggested. "No," he said. Then he tried to explain to me how the emotional release of the tears was something else.

10th Mountain Division commander, Gen. George P. Hays, uses a microphone to announce to the troops on Lake Garda that the war in Italy has been successfully concluded by unconditional surrender on the part of the Germans. May 2, 1945. (*Denver Public Library*).

The soldiers had been plucked from civilian life, he explained, were trained, became part of a system, learned a routine. Then they were brought to a country they had never expected to visit. They had lived like animals in the field, saw horrors they never thought possible, lived with dead bodies and with the likelihood their own lives would end, or that they would be disfigured or maimed. They had survived combat fatigue, the constant strain of being at the mercy of fate, under almost constant threat of enemy artillery and mortar fire. They had been part of history. Their lives had meaning and purpose. They were fighting for their country, for their way of life. In less than three weeks, they had come out of the Apennines and fought their way across the Po Valley to the Alps. They were always on the move. They did everything that could be asked of a man. Every day was new and exciting and filled them with pride. They were important and successful in a way they had never experienced in civilian life.

If necessary, they were prepared to die.

And now all that was quite suddenly, unexpectedly—over. They had survived. Certainly they were relieved to be alive. But for what? To return to some ho-hum job in the small town they came from?

It was not joy that brought the tears. What Dad told me was that some of the men who cried were, in a way, sorry to see the war end! I cannot say that was his feeling, as well, but he seemed to understand why some of the soldiers cried when they heard the war was over. He was certainly able to sympathize with them.

In a way, life—as they had come to know it—was over.

49

UNCONDITIONAL SURRENDER

Most wars end with an armistice followed by a negotiated peace treaty. Undoubtedly, that is what Hitler envisaged at the outset of the war. If he lost, he expected to be exiled, as was Napoleon, *twice*, and as was the Kaiser after World War I. *Unconditional surrender*, insisted upon by President Roosevelt in January 1943—without consulting his military advisors—was almost unheard of. In the aftermath of World War I, Adolf Hitler had come to power claiming Germany did not lose World War I militarily, but had been "stabbed in the back" at home by Socialists, Bolsheviks, Jews, and corrupt Social Democrats. This time FDR wanted Germany to have no such excuses. But "unconditional surrender" created issues.

Before 1945, German prisoners were generally shipped to prisoner-of-war ("POW") camps in the United States on the returning, otherwise empty ships that had brought American soldiers to the front. In fact, by 1945, over 350,000 German and 50,000 Italian POWs were in hundreds of camps in the United States. The collapse of the German army in Italy in April–May of 1945 meant hundreds of thousands of new POWs for the Allies to deal with, plus their equipment, vehicles, arms, etc. Would the POWs be sent to the USA? And what of all the stolen art? The surrender documents in Caserta covered some of the anticipated issues, but certainly not all.

Fifteenth Army Group Commander General Mark Clark invited German General von Vietinghoff to come from Bolzano to his headquarters in Florence with staff members for a formal surrender and to discuss with his own staff details of implementing the cessation of hostilities in Italy. However, partisans in the Bolzano area had attacked von Vietinghoff's headquarters May 2, and he did not want to travel through hostile territory; so Clark asked him to send one of his generals. Von Vietinghoff agreed to send Oxford-educated General Fridolin von Senger und Etterlin, who spoke perfect English and had masterminded the German defenses at Monte Cassino. A staff member of the German party was the sophisticated SS Colonel Eugen Dollmann, the man who seemed to know everyone, who had in the past translated for Hitler and Mussolini, and who participated in the Operation Sunrise meetings in Switzerland. Another, was Lieutenant Colonel Viktor von Schweinitz, who had signed at Casserta on behalf of General von Vietinghoff

Down the Allied chain of command, 10th Mountain Division's General Hays ordered General Ruffner, his artillery chief, to head for Bolzano in the Alps in a caravan of four jeeps to meet with General von Vietinghoff and bring General von Senger and the German staff members back with him. On the way back to General Hays's command post on Lake Garda, they had to take DUKWs around the still damaged road and tunnels on

German Gen. Fridolin von Senger und Etterlin (left) with 10th Mountain Division commander Gen. George P. Hays and Fifth Army chief of staff Gen. Don E. Carleton in Verona May 3, 1945. (*Signal Corps*).[c]

the east shore of the lake, that had held up Dad's battery for so many days, to General Hays's command post on southern part of the lake. After eating dinner in a tent, they stayed at a hotel in Verona for the night.

While the German party was on its way, an exciting thing happened back at their headquarters in Bolzano.

SS General Wolff got a phone call from Field Marshal Kesselring, now in Western Austria. Kesselring asked Wolff to use Little Wally to radio Caserta about whom to contact for *surrender of Army Group* G, which constituted the German armies in Austria and Southern Germany.

The surrender in Italy had placed the German units north of Italy in an impossible military situation. Field Marshal Alexander radioed back that the question was referred to General Eisenhower. The next day, detailed instructions were sent from General Eisenhower's command in France to Field Marshal Alexander in Casserta to Little Wally and General Wolff in Bolzano, and by phone from there to Field Marshal Kesselring—for how Kesselring, or a representative, could approach the American lines to discuss surrender. The Caserta surrender became a template for ending the war in Europe.

According to his memoir, General Truscott, Fifth Army commander, whose headquarters was in an olive grove near Verona, had his chief of staff, General Carleton, meet and make arrangements on May 3 for the German surrender party in Italy. General Truscott, who had "no desire to exchange civilities" with enemy generals, did not meet with the Germans. The next morning Truscott and the Germans flew separately to General Mark W. Clark's 15th Army Group headquarters at Florence for a formal surrender ceremony. General Truscott, not one for ceremony, had no desire to be there, just as he had no desire to swap war stories with the Germans, who were driven to the headquarters that morning in Allied staff cars from the nearby airfield

Prior to the brief ceremony in General Clark's office, the Allied generals had removed their side arms, stacking their pistols and belts on a table in the corner. At about 10:30 a.m., German General von Senger, obviously uncomfortable, arrived in a dark green uniform with an Iron Cross around his neck and a small pistol in a holster behind him. He accidentally stepped on General Clark's cocker spaniel, Pal, who yelped—adding to the awkwardness of the situation.

In official surrender ceremony, German Gen. Fridolin von Senger und Etterlin and 15th Army Group commanding general Mark W. Clark at Clark's spartan office near Florence, Italy. (Note Gen. von Senger was wearing a sidearm behind him while the Allied officers were unarmed.) To the left of Gen. Clark in the picture is Eighth Army commander British Gen. Richard L. McCreery. Obscured is Maj. Gen. Benjamin W. Chidlaw of Twelfth Air Force. At extreme right in photo, in his traditional leather jacket, is Fifth Army commander Gen. Lucian K. Truscott, Jr., obviously not wanting to be there. (He later called the surrender ceremony "pointless" except as "a photographic record in the Hollywood tradition."). May 4, 1945. (*Imperial War Museums*).[c]

While the shutters of the still cameras clicked away and the newsreel cameras whirred, General von Senger gave the traditional German Army salute—no longer having to give the stiff-arm "Heil, Hitler"—and General Clark saluted back. Von Senger then said, in English, he was reporting on behalf of General von Vietinghoff for instructions for the surrender of Army Group C. Whereupon, General Clark handed General von Senger written instructions and told him his chief of staff, General Gruenther, would preside over a conference of the respective Allied and German staff officers in a nearby modular building to iron out details of the surrender.

Von Senger saluted, did an about-face, and walked out. As the German general was leaving, General Clark first noticed the holster von Senger was wearing behind him. A reporter said, "General, you let that guy surrender while armed!" General Clark told a guard to bring von Senger back in. When the German general returned, a quizzical look on his face, General Clark said, "Get rid of that gun." Von Senger went out the door again and unbuckled his leather belt and holster, dropping them by a tree. To General Clark, it was a metaphor for the disarming of the enemy, signifying the fighting in Italy truly was over.

He then ordered an aide to fetch the weapon, which became a prized possession in General Clark's souvenir collection.

50

RESIA

While the "photo-opportunity" surrender was occurring in Florence the morning of May 4, 1945, Dad's battery was on the move. The engineers had finally repaired the road and tunnels on the east shore of Lake Garda, so Dad's battery, stuck since April 28, could finally get to the north end of the lake. Dad's battery had been chosen to be part of another 10th Mountain Division task force, this time with the mission of going through what had been enemy territory only two days before—all the way to the Austrian border. Battery C also participated. The long convoy actually was comprised of *two* task forces, one from the 86th Mountain Infantry Regiment and one from the 87th, plus armor and artillery support, plus numerous trucks, jeeps, and other vehicles. It went east from Torbole, at the north end of Lake Garda, toward Rovereto, then north on the main highway leading to Trent, Bolzano, and the Brenner Pass, gateway to Austria.

Albert H. Meinke, Jr., M.D., an army doctor for the 86th Mountain Infantry, was in the same convoy with Dad. He wrote about the trip in a memoir called *Mountain Troops and Medics* (1993). He described the eerie sensation of riding past large groups of German soldiers watching the convoy as it passed by. The Germans, in what had been their rear, were described as neat, well-dressed, disciplined, erect, and proud—in contrast to the many dejected prisoners taken in the recent campaign. The soldiers by the side of the road were more or less expressionless. The convoy passed a number of German vehicles traveling in the opposite direction, tanks and trucks with towed artillery pieces, which made the Americans apprehensive; but there were no incidents of hostility.

As the convoy approached Trent, traffic got congested, and German military police guided the way. Italians cheered the Allies on but were reserved whenever German soldiers were nearby. Dad, from his kitchen truck, would have seen Trent's railroad yard, which had been completely destroyed by Allied bombs.

From Trent, the convoy went north toward Bolzano, German army headquarters in the South Tyrol German-speaking part of the Italian Alps. The closer the convoy got to Bolzano, the worse the traffic congestion. German soldiers with rifles across their backs were directing traffic at almost every intersection. Walking in the opposite direction were many freed Italians who had been used as slave laborers by the Germans. Mixed with them were Czechoslovakian soldiers drafted into the German army, happy the war, for them at least, was over.

Dr. Meinke described the convoy as being *eight miles long*. The air got colder as the sun went down. Scattered snowflakes fell as the Italians in Bolzano cheered at the sight of the Allied convoy.

After passing through Bolzano, the convoy did not head northeast toward the Brenner Pass, as some had expected, but instead headed northwest in the darkness toward Merano. The higher elevation brought colder air and more snow as the convoy proceeded in usual blackout conditions. Then, orders came down; the blackout was no longer necessary. The trucks turned on their headlights in the falling snow—creating a beautiful sight. It was another sign to Dad and others in the convoy that the war—in Italy, at least—was really over.

At about midnight, the convoy passed through Merano, noted for its hospitals and near where a lot of stolen Italian art was hidden. The miles of headlights from the vehicles in the convoy must have alerted the citizens of Merano the Allies had arrived, because many people came out to wave, despite the late hour. From Merano, the convoy kept climbing. The snow continued.

At about 5 a.m. on May 5, the convoy halted at the Austrian border at

Lago di Resia (Reschensee) as it looks today. The bell tower of Curon (Graun), dating from 1357, survived when the town was flooded as part of a hydroelectric project in 1950. (Noclador)

the Resia (*Reschen*, in German) Pass, at about 5,000 ft. in elevation, the only pass between Italy and Austria west of the Brenner. German artillery on the Austrian side of the border had the convoy in its sights as the task force officer in charge met with the German commander in Nauders, just over the border. The German army in Austria, part of Army Group G, had not yet surrendered. The American commander persuaded the German officers the task force was not intended to attack the German forces in Austria, but to prevent German units from entering Italy, and to prevent any rogue German units in Italy from escaping into Austria. The Germans understood their army in Italy had surrendered, and the Allies had a right to stay on the Italian side of the border. Later that day, German Army Group G actually did surrender at Haar, just outside Munich, in Bavaria—thanks in part to Little Wally—effective the following day.

Dad spent about a week in the vicinity of Resia, which was at that time an Alpine resort area in a valley that contained three lakes. After the war, the valley was flooded as part of a hydroelectric project and the three lakes became one big lake known as Lago di Resia.

It was here in the snowy Italian Alps, right near the Austrian border, that Dad was stationed on May 8, 1945, the effective date of the final German surrender in Europe, celebrated as the date of final victory in Europe, or "V-E Day."

51

GUARDING

With the war in Europe now over, I am sure Dad wanted to go home right away. But so did millions of other American soldiers, sailors, and airmen who had been brought to Europe over several years. With the war still going on in the Pacific and the draft still in place, the War Department had a huge task deciding how best to utilize the millions of men (and women) in Europe who would not be able to go home for many months.

Dad was soon aware, despite the end of the war in Europe, his sojourn in Italy was far from over.

On May 12, 1945, Dad's battalion was released from its long attachment to the 10th Mountain Division. In 114 days of combat, the 10th Mountain had suffered nearly one thousand killed and four thousand wounded. Because of its success in Italy, the division, as a unit, was selected to return to the States to prepare for the anticipated invasion of Japan.

Dad's battalion was attached to the 428th Field Artillery Group for operational control.

On May 13, Dad's battery left Resia by motor convoy in the afternoon and arrived in the vicinity of Merano about 5:15 p.m. for the night. The next morning, Dad's unit left Merano for the vicinity of Feltre and Belluno, Italy, in a picturesque part of the Alps called the Dolomites, east of Trent and northwest of Venice, with a new assignment: guarding captured medical and engineering ordnance installations. The M7s, with their .50-caliber machine guns, were ideal for guard duty. On May 22nd, Dad's battery was moved to Brunico (*Bruneck*), Italy, northeast of Bolzano, to guard enemy signal and ordnance dumps in the area. A week later, they guarded similar captured installations near Bressanone (*Brixon*) closer to Bolzano.

Guard duty was quite boring. The Germans and Fascists had to give up their weapons, of course. Surprisingly, the partisans mostly complied with the demands of the Allied military government and gave up their arms, as well. Such arms ended up in ordnance dumps like those guarded by Dad's battery. I remember Dad telling me some of the bored men in his unit would sit on piles of captured weapons and fire bursts from

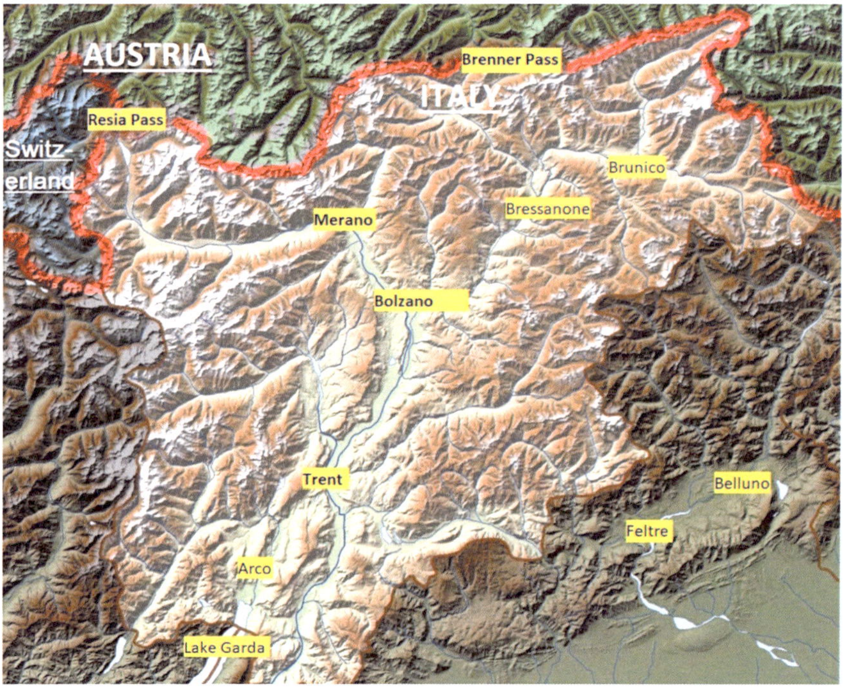

The Northern Italy area, mostly in the Alps, northwest of Venice, where Dad's battery (A) did guard duty.

Dad brought home this photo (originally in color, now faded) processed June 28, 1945. Not the clearest of pictures. I think Dad is the sergeant standing in front of the M7.[c]

German submachine guns, called "burp" guns. The men would reload the burp gun and continue firing until the barrel got hot, then would throw down the weapon, pick up another, and continue shooting.

When Dad's battery left the pass at Resia on May 13, Battery C went to Arco, just north of Lake Garda where Battery B and the headquarters and service batteries of the battalion were located and had assumed military control of the town. But soon Batteries B and C were sent out to guard captured enemy ordnance and supply dumps, too—while battalion headquarters remained at a villa in Arco.

On June, 11, 1945, Dad's battery was relieved of its special guard duties at Bressanone and went by motor convoy south to Arco, battalion headquarters, just north of Lake Garda.

Leonard Warner, who was in the headquarters battery of Dad's battalion, said in his memoir *Metamorphosis in the Mediterranean* the headquarters battery built a fancy bar and dance hall in the solarium of the sanitarium where they were billeted in Arco. The Americans had discovered a German stockpile of beers, wines, whiskeys, and other liquors in a warehouse about two miles out of town. Mr. Warner called it "a veritable treasure store of booze." The nightly partying and dancing by the soldiers supposedly attracted a lot of the young Italian women of the area. In short, Arco was a fun place after the war for the 1125th Armored Field Artillery Battalion.

On June 24, 1945, Dad's battery left Arco for a place called Villa

Nuova, which I assume was somewhere in the Lake Garda/Verona area, where it guarded quartermaster and engineering installations until July 4.

Dad told me about how he frequently was amused by the GI's favorite inside joke. Wherever he went in Italy, and elsewhere during his time in service—on walls, on buildings, in ports, on airplane hangars, in and on bivouacs, on vehicles, on ships, in latrines—seemingly everywhere—Dad saw the omnipresent GI graffiti, "KILROY WAS HERE," often accompanied by a simple drawing supposedly showing the legendary, mysterious, ubiquitous, mythical Kilroy peering over a wall.

At noon on the Fourth of July, 1945, Dad's battery fired a 48-gun salute (one for each state at the time?) at Fifth Army headquarters, presumably in the Verona area. In the afternoon, they went by motor convoy to Pacengo, Italy near the southeast shore of Lake Garda. Dad was still responsible for the mess for about a hundred men.

The inescapable Kilroy.

52

LOOSE TIME

I cannot account for all of Dad's time while he was in Italy after the fighting ended. I am sure he visited a number of major Italian cities. Fifth Army encouraged the soldiers to partake in cultural and educational programs, and I assume leave was not hard to get. I know he visited Milan and Rome at some point, and likely went to a number of other Italian cities. He may occasionally have connected with friends from his various postings.

Dad was most likely shocked, as were most Americans, that in a July, 1945 election in Britain, the conservatives lost in a landslide; and wartime leader Winston Churchill had to resign as prime minister. Churchill's visit to Lake Como in Italy several weeks later for a "painting holiday" fueled speculation the former prime minister had ordered the death of Mussolini and was trying to recover incriminating correspondence he had sent to the Italian dictator.

One of Dad's favorite pastimes in the service was *reading*. The soldiers had available a newspaper called *Stars and Stripes* and the monthy magazine *Yank* with its pinup girls. In 1991 Dad and Mom visited us in Santa Monica, California; and actress Yvonne de Carlo, then close to seventy, was seated near us at a restaurant. When I pointed her out to Dad, he said, "She sure didn't look like that in the '40s!"

Yank featured Sergeant George Baker's cartoon character "The Sad Sack" while *Stars and Stripes* had the "Willie and Joe" cartoons of Sergeant Bill Mauldin, popular with all the GIs in Italy. Mauldin, who was wounded at Monte Cassino, was awarded the Pulitzer Prize for his cartoons in 1945.

PXs carried newsmagazines like *Look*, *Time* and *Life*. Dad read a lot of novels, a habit that carried over to when he became a civilian. I remember mornings he admitted getting little sleep because he stayed up late reading a page-turner. Hollywood movies, of course, were also available to entertain the soldiers in Italy. Like most American GIs at the time, I am sure Dad wondered what was going to happen to him. How soon before he could go home? What would he do while waiting to go home? Would he eventually be shipped off to the Pacific to fight against Japan?

I do not know where or when this photo of Dad was taken. Since the first sergeant, in the middle, and the sergeant, on the right, have Fifth Army patches, I assume somewhere in Italy.

Actress Yvonne De Carlo as a *Yank* pinup in 1944, and (insert) around when Dad saw her.

July brought good news to the city of Florence. Thanks to information provided by SS General Karl Wolff, the Allies were able to locate much of the art that had been stolen by the Nazis and hidden in the Alps around Merano and Bolzano. On July 22, 1945, a convoy of Fifth Army trucks returned most of the priceless art stolen from Florence by the Nazis the previous year. If Dad made it back to Florence after that time, he might have been able to see the museum art he missed on his previous visit to the Florence rest center.

On July 28, 1945, in thick fog over New York City, a twin-engine B-25

George Baker's "The Sad Sack" and Bill Mauldin's "Willie and Joe."

Fifth Army trucks entering Piazza della Signoria in Florence, Italy near the Uffizi Gallery, July 22, 1945 with art works looted by the Germans. (*National Archives*).[c]

Mitchell bomber, on a routine mission, accidentally flew into the 79th floor of the tallest structure in Manhattan, the Empire State Building. All three men aboard were killed, as well as eleven civilians in the building.

Most Americans do not appreciate the development of the country's awesome World War II air power came at a heavy cost in human lives. An estimated thirteen to fifteen thousand American service personnel were killed *in the US* during the War in over six thousand fatal crashes of military aircraft.

53

THE POINT SYSTEM

Dad, no doubt, saw a War Department propaganda film called *Two Down and One to Go*, about the military's changing manpower needs. Italy and Germany were out of the war, but a major buildup was necessary in the Pacific. Although some units in the Mediterranean Theater—like the 10th Mountain Division, chosen for reassignment to the Pacific Theater—left Italy intact, most units were destined for *deactivation*, with their personnel returning to the United States *individually* under a priority system called the Advanced Service Rating Score (ASRS). The higher the score, the sooner the soldier could expect to go home. An enlisted man earned one point (or credit) for each month *in service* and an additional point for each month served *overseas*. In addition, five points were given for each decoration and campaign bronze battle-star earned.

If you recall, Dad was slightly injured when he was buried alive in his foxhole by an exploding mortar round, but refused to go to the battalion aid station and possibly qualify for a Purple Heart Medal. After the fighting ended, he wished he would have gone after that medal because it would have made his ASRS score five points higher, meaning he could have come home sooner. Soldiers also got twelve points for each child under eighteen—which did not help Dad. When the system was first implemented in May 1945, a soldier needed at least 85 points to go back to the States.

On August 2, 1945, while Dad was still at Pacengo, an order came down transferring him and a number of other enlisted men from the 1125th—which he had been in since it was formed—to the 536th Field Artillery Battalion. Dad was assigned to Battery A, as he had been in his old battalion. Apparently much of the equipment, likely including Dad's kitchen truck, remained in Pacengo, while the remains of the 1125th Armored Field Artillery Battalion headed for Florence and eventual deactivation about five weeks later. According to the order, Dad's ASRS score was only **58**. At two points per month (one for time in service and one for time overseas), Dad anticipated being in Italy for quite a while—though, eventually, the minimum score for going home was likely to decrease.

On August 23, Dad's new battery moved from Pacengo to Verona. The same day, Dad was assigned to temporary duty at a new far-away post in *Southern Italy* that he really loved.

54

FOGGIA

Dad's temporary assignment was at the Adriatic Base Command in Foggia, Italy. The relatively large city of Foggia in the region of Apulia (Puglia) was about ninety miles northeast of Naples, across the mountains. It was about twenty-five miles inland from the Adriatic Sea. Situated on a fertile plain, Foggia was sometimes called the "granary of Italy." The flat land in the area made it ideal for use by the Italian Fascist air force, which built a number of airfields in the vicinity. In September of 1943, when Mussolini fell from power, the Germans briefly took over the airfields, which were heavily bombed by the Allies. Soon the airfields were captured by the English-led Eighth Army and were repaired and used by the Allies, especially for strategic bombing. Additional airfields were also built. An estimated *two dozen* airfields were carved out of the farmland within twenty-five miles of Foggia. As I understand it, the Adriatic Base Command, created earlier in 1945, supported the Twelfth and Fifteenth Air Forces, primarily with hospital and medical assistance.

One of the goals of the War Department during this period was to try to give idle soldiers skills that might be useful back in the States,

Southern Italy. Dad managed a PX in Foggia.

to make the transition to civilian life a little easier. The *United States Armed Forces Institute*, headquartered at Madison in Dad's home state of Wisconsin, offered dozens of "education manuals" to help soldiers learn useful skills on their own, such as electricity, auto mechanics, plumbing, shorthand, agriculture, foreign language, and business, among many others. In addition, a number of educational courses and training programs were offered to the GIs. One was in Foggia.

Dad was *not* a mess sergeant in Foggia. He told me he *managed a PX* there. His records say he attended a "store manager" service school in 1945, and I assume the training he received was in Foggia. When Dad worked at the A&P back in West Allis he dreamed about being a grocery store manager some day; and when a store manager training course was offered by the Army, he signed up.

The PX or Post Exchange was basically a retail store operated by the War Department under the Army Exchange Service. The PX sold goods and services not otherwise provided to the ordinary GI Joe or Jane through the military supply chain, such as tobacco products—cigarettes for a nickel a pack—toiletries, candies, wallets, watches, stationery, board games, reading material, military clothing, and nylon stockings (popular with the Italian girlfriends), to name only a few.

While the management of the PX at that time was military, most of the employees were local civilians. Dad's ability with Italian, limited as

Dad brought home this picture of the PX he managed in Foggia, Italy.[c]

it was, came in handy. He had to confront personnel problems, order and keep track of inventory, balance the books, etc. In short, he learned how to manage a store, something that would come in handy later on in his life. He took advantage of his time in Foggia and was able to transfer leadership skills he had learned as a mess sergeant to the management of the PX.

On August 6, 1945, America's top flying ace, Major Richard I. Bong of Poplar, Wisconsin—with forty victories over the Japanese in a twin-engine P-38 Lightning to his credit—was killed in North Hollywood, California testing a new Lockheed P-80 jet aircraft.

The same day, the *Atomic Age* began, as Americans dropped an enriched *uranium-235* bomb on Hiroshima, Japan. Three days later, a *plutonium* bomb was dropped on Nagasaki.

In 1943, while Dad was at Camp Davis in North Carolina, the Corps of Engineers, back in Wisconsin, constructed a special two-block long plant on the east end of the Allis-Chalmers property in West Allis, facing Hawley Road, for the *Manhattan Project*. The high-security Hawley Plant was where Allis-Chalmers secretly manufactured over 940 huge (14-ton) magnetic coils, as well as conductors, made with US Treasury silver—due to its high conductivity and the wartime scarcity of copper—for use at Oak Ridge, Tennessee, where three different methods were used to make

Maj. Richard I. Bong America's top ace.[c]

the 140 lbs. of uranium-235 for the Hiroshima bomb. Allis-Chalmers made several other components for the Manhattan Project, as well. (The plutonium-239 in the Nagasaki bomb was made at the rural so-called *Hanford Site* outside of Richland, Washington).

Fleet Admiral William D. Leahy, who had grown up in Ashland, Wisconsin and was *de facto* Chairman of the Joint Chiefs and top military advisor to President Truman—as he had been to President Roosevelt—opposed dropping the atomic bomb on grounds its radiation effects made it barbarous, like poison gas. He argued the naval embargo of Japan and conventional bombing had already brought the Japanese to the point of surrender. President Truman, however, wanted to shorten the War and save American lives—at any cost.

On August 14 (August 15 in Japan), while Dad was still up north in Pacengo—in an unprecedented event—Japanese Emperor Hirohito went on the radio to announce to his subjects Japan had agreed to an armistice and unconditional surrender. The fighting in the Pacific was finally over—meaning *World War II was over.*

On September 2, 1945, while Dad was in Foggia, a formal surrender ceremony occurred on the USS Missouri in Tokyo Bay, presided over by Five-Star General (and former Milwaukee resident) Douglas MacArthur.

That same day, Dad's ASRS score was calculated at **65**. He got 31 points for months in the Army, plus 19 more for months overseas. Finally,

Dad's Fifth Army patch.

he got 15 points combat credit, five for each of the three campaign bronze battle-stars he earned: *Rome-Arno, Northern Apennines*, and *Po Valley*. Fortunately, as time went on, fewer and fewer than 85 points were needed for going home.

On September 9, Fifth Army became nonoperational, having turned over responsibility for its units and personnel to the Mediterranean Theater for redeployment, demobilization, or return to the United States.

Fifth Army was the first American army ever created outside of the United States, having been officially activated January 5, 1943 at General Clark's headquarters in Oujda, French Morocco. Its original mission was to defend French Morocco and Algeria and train troops for the invasion of Italy.

Its shoulder patch had a red background with the blue silhouette of a Moroccan mosque, a large white letter "A" and a smaller white "5" below it.

Since the 1125th Armored Field Artillery Battalion was an independent Fifth Army unit, I think the Fifth Army patch was the only unit patch Dad wore during the fighting in Italy. Fifth Army was officially deactivated on October 2, 1945 at Camp Myles Standish, Massachusetts (only to be reactivated the following year).

On September 11, 1945, Dad's old unit, the 1125th Armored Field Artillery Battalion, was officially disbanded.

The same day, Dad, still at the PX in Foggia, celebrated his 21st birthday.

55

NAPLES, AGAIN

The Adriatic Base Command was scheduled to close down at the end of October 1945. On October 20, Dad left Foggia and was returned to duty with Battery A of the 536th Field Artillery Battalion, which had moved from Verona down to Naples. Presumably, he was back to being a mess sergeant. A week later, he was assigned by order of the 8th Port of Embarkation (in Leghorn) to the 3334th Quartermaster Truck Company of the 76th Quartermaster Battalion (Mobile), in Naples. The Port of Embarkation apparently put him in that unit to prepare to ship him back to the States from Leghorn. With Thanksgiving only a month away, I am sure Dad was wondering if he would make it home for Thanksgiving.

On the morning of November 9, Dad left Naples by rail destined for the port of Leghorn in Tuscany, near Florence, and arrived the following morning. I am sure Dad believed he would be sailing out of Leghorn for home any day. But, of course, he had to wait his turn.

Three days after he arrived in Leghorn, Dad received the order he was waiting for, dated November 13. He finally qualified to go home—

but instead of sailing out of Leghorn, he was ordered back south to Naples. He was to report to the Leghorn railroad station November 16 to return by train to Naples and report to the 7th Replacement Depot in Naples for processing. Dad had to spend Thanksgiving in Naples, but he had high hopes he would be back in West Allis for Christmas.

56

HEADING HOME

With the War now over, even in the Pacific, the War Department was making every effort to bring overseas service personnel home by Christmas. The plan was known as *Operation Magic Carpet,* and it involved, among other things, bringing several aircraft carriers from the Pacific through the Panama Canal to be refitted as troop transports on the Atlantic. The *U.S.S. Wasp* (CV-18), ninth vessel to bear the name, one of the most celebrated warships in the Pacific Theater, was such a ship.

During the first two weeks of November, 1945, at the New York Naval Yard, the *Wasp,* one of 24 Essex-class carriers built during the War, was relieved of its planes, ordnance, etc.—and its huge hangar deck was refitted with bunks, cots, hammocks, and bedrolls to carry troops. On November 15, the *Wasp* sailed the short distance to a prisoner of war camp at Bayonne, New Jersey, picked up about 1,200 Italian prisoners of war, and headed for Naples, Italy.

According to one account, the *Wasp* needed only twenty-seven hours in the port of Naples to deliver the Italian POWs, refuel, resupply, and

The U.S.S. *Wasp* (CV-18), the ship that brought Dad back to the States. (*U.S. Navy*).[c]

take aboard thousands of GIs eager to go home, Dad among them. The *Wasp* left Naples on November 27, 1945.

In processing out, Dad declared he had certain property he had actually purchased (probably from his PX) and that was his *personal* property: a "val pack" (canvas garment bag), khaki shirt, 2 ties, a pair of tank boots, and a belt with brass buckle. He also declared a pistol as supposedly captured from the enemy. However, the pistol was actually an American-made .38 cal. Harrington & Richardson. Dad told me he brought a gun home from the war and got rid of it when I was born. Mom remembers it as a revolver. Another souvenir brought home by Dad was a medal (without ribbon) he took off a dead German. It was a *West Wall Medal* awarded for service defending Germany's western border.

When boarding the *Wasp*, Dad was given a booklet of "passenger regulations" that contained a welcome from Captain Wendell G. Switzer who promised three meals a day, movies in the forward part of the Hanger Deck, and a library. The Flight Deck was available for exercise. With the *Wasp* having so many more men on it than it was built to accommodate, everyone was asked to conserve fresh water as much as possible—only *salt water* showers available.

Hanger deck on the Wasp during *Operation Magic Carpet*. (*U.S. Navy*).[5]

Despite the cramped quarters, there were few complaints. Everyone was anxious to get home. I remember Dad telling me the aircraft carrier was more stable and much faster than the Liberty Ship he went overseas on. He went through the Mediterranean Sea and crossed the Atlantic in only a week, or so. On December 4 (Dad's military records) or 5 (a newspaper account), 1945, the *Wasp* docked at the Hampton Roads Port of Embarkation in Newport News, Virginia, with 5,656 aboard. All aboard were excited to see America again, many having been away for several years. A big sign had been painted on the dock at Newport News that read "Welcome Home, Well Done."

The returning service personnel walking down the gangways, each

Entrance to Camp Patrick Henry in 1945. (*Signal Corps*).

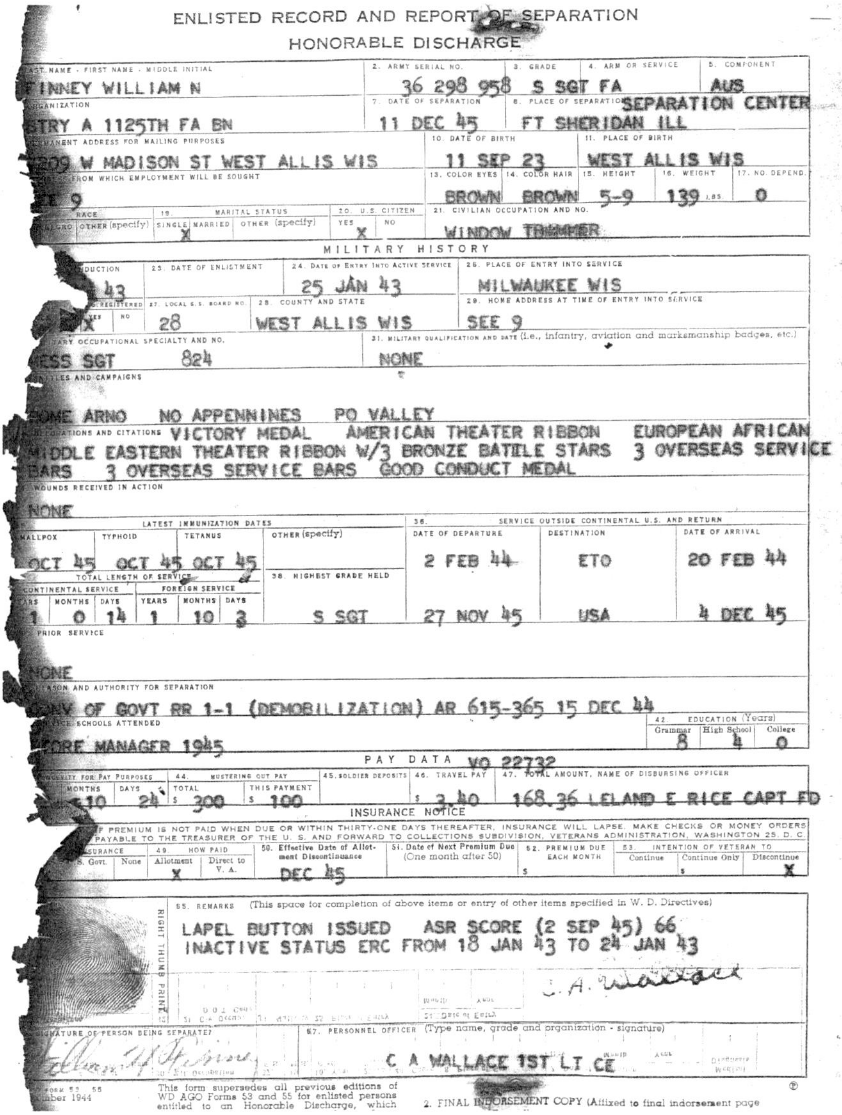

ENLISTED RECORD AND REPORT OF SEPARATION
HONORABLE DISCHARGE

Last name - First name - Middle initial	2. Army serial no.	3. Grade	4. Arm or service	5. Component
FINNEY WILLIAM N	36 298 958	S SGT	FA	AUS

Organization	7. Date of separation	8. Place of separation
BTRY A 1125TH FA BN	11 DEC 45	SEPARATION CENTER FT SHERIDAN ILL

Permanent address for mailing purposes	10. Date of birth	11. Place of birth
2209 W MADISON ST WEST ALLIS WIS	11 SEP 23	WEST ALLIS WIS

Address from which employment will be sought	13. Color eyes	14. Color hair	15. Height	16. Weight	17. No. depend.
SEE 9	BROWN	BROWN	5-9	139 LBS.	0

Race	19. Marital status	20. U.S. citizen	21. Civilian occupation and no.
	SINGLE X	YES X	WINDOW TRIMMER

MILITARY HISTORY

Induction	23. Date of enlistment	24. Date of entry into active service	25. Place of entry into service
43		25 JAN 43	MILWAUKEE WIS

Registered	27. Local S.S. board no.	28. County and state	29. Home address at time of entry into service
YES X	28	WEST ALLIS WIS	SEE 9

Military occupational specialty and no.	31. Military qualification and date (i.e., infantry, aviation and marksmanship badges, etc.)
MESS SGT 824	NONE

Battles and campaigns: ROME ARNO NO APPENNINES PO VALLEY

Decorations and citations: VICTORY MEDAL AMERICAN THEATER RIBBON EUROPEAN AFRICAN MIDDLE EASTERN THEATER RIBBON W/3 BRONZE BATTLE STARS 3 OVERSEAS SERVICE BARS 3 OVERSEAS SERVICE BARS GOOD CONDUCT MEDAL

Wounds received in action: NONE

Latest immunization dates: Smallpox	Typhoid	Tetanus	Other (specify)
OCT 45	OCT 45	OCT 45	

36. Service outside continental U.S. and return: Date of departure	Destination	Date of arrival
2 FEB 44	ETO	20 FEB 44
27 NOV 45	USA	4 DEC 45

Total length of service: Continental service Years	Months	Days	Foreign service Years	Months	Days	38. Highest grade held
1	0	14	1	10	3	S SGT

Prior service: NONE

Reason and authority for separation: CONV OF GOVT RR 1-1 (DEMOBILIZATION) AR 615-365 15 DEC 44

Service schools attended: STORE MANAGER 1945

42. Education (Years): Grammar	High School	College
8	4	0

PAY DATA VO 22732

Longevity for pay purposes: Months	Days	44. Mustering out pay: Total	This payment	45. Soldier deposits	46. Travel pay	47. Total amount, name of disbursing officer
10	24	$ 300	$ 100		$ 3.40	168.36 LELAND E RICE CAPT FD

INSURANCE NOTICE

IF PREMIUM IS NOT PAID WHEN DUE OR WITHIN THIRTY-ONE DAYS THEREAFTER, INSURANCE WILL LAPSE. MAKE CHECKS OR MONEY ORDERS PAYABLE TO THE TREASURER OF THE U. S. AND FORWARD TO COLLECTIONS SUBDIVISION, VETERANS ADMINISTRATION, WASHINGTON 25, D. C.

Insurance: U.S. Govt.	None	49. How paid: Allotment	Direct to V. A.	50. Effective date of allotment discontinuance	51. Date of next premium due (One month after 50)	52. Premium due each month	53. Intention of veteran to: Continue	Continue only	Discontinue
		X		DEC 45		$		$	X

55. REMARKS (This space for completion of above items or entry of other items specified in W. D. Directives)

LAPEL BUTTON ISSUED ASR SCORE (2 SEP 45) 66
INACTIVE STATUS ERC FROM 18 JAN 43 TO 24 JAN 43

RIGHT THUMB PRINT

C. A. Wallace

Signature of person being separated

57. PERSONNEL OFFICER (Type name, grade and organization - signature)
C A WALLACE 1ST LT CE

WD AGO Form 53-55, November 1944. This form supersedes all previous editions of WD AGO Forms 53 and 55 for enlisted persons entitled to an Honorable Discharge, which
2. FINAL INDORSEMENT COPY (Affixed to final indorsement page)

Dad's restored separation document, partially burned, like the rest of his records, in the 1973 fire at the National Personnel Records Center in St. Louis. It gave his year of birth as 1923—whereas he was really born in 1924. Dad died in 1995 not knowing this and his other military records, though damaged, survived the fire.

carrying a duffle bag, were serenaded by the port band playing in the dock area.

Twin train tracks ran directly onto each of the ten piers in Newport News. The Army's Camp Patrick Henry, where Dad had previously stayed while awaiting shipment overseas, was a short train ride away. The main entrance to the camp was under a large arch that now said "Welcome Home." With the War over, the post had become a festive place, the camp band playing almost continually, like at the dock.

Dad would have stayed in the camp's tarpaper-covered barracks for a few days to get his "land legs" and for processing. He could have gone to the movie theaters, service clubs, recreation halls, general store, and "Last Chance" night club on post and would have had one of the best steak dinners of his life, given to all the returning GIs. He likely used one of Camp Patrick Henry's telephone call centers to let Uncle Bill know he was back in the USA and would be home soon. Within a few days, probably with a fresh haircut, he was in his dress uniform and on his way by rail to Chicago, and then Fort Sheridan, for what was called "demobilization separation."

According to Dad's Form WD 53-55 (the equivalent of today's DD-214), his separation from active duty occurred on **December 11, 1945** at the U.S. Army Separation Center at Fort Sheridan, Illinois. Army chow and the passage of time had beefed him up. He weighed 139, sixteen pounds more than when he was inducted nearly three years before, and he had grown an inch taller.

I do not know how he got home from Fort Sheridan. Uncle Bill probably drove him the seventy miles to West Allis, or he could have gone by bus or from either of Fort Sheridan's two train stations: one for the *Chicago & Northwestern* and the other for the electric interurban *North Shore* that ran between the Chicago Loop and downtown Milwaukee.

57

BACK HOME

The holiday season in 1945 was especially joyous with the war over and many of the GIs home for the holidays. The great thing for Dad was seeing his family: Uncle Bill, little Jill, his cousin Kay, her husband Ray, Kay's sisters Gertrude and Margaret, and, of course, his institutionalized mother. He was also able to get together with his old friends in West Allis, some of whom he had known since kindergarten. They all had been in the service themselves and had stories to tell about their travels and adventures during the War. They decided one day to pose for a photo in their uniforms.

Dad was now twenty-one and could legally drink. As I mentioned, he got a driver's license. He soon bought a car. Dad had sent home an allotment into savings all the while he was in the service, so he had quite a bit saved up, supplemented by his gambling winnings. He also told me he found some foreign money on a dead German, and his commanding officer told him to keep it. I suspect he traded it, probably on the black market, for Allied Military currency, which he converted to greenbacks when he left Italy, or he could have exchanged it at Camp Patrick Henry. Mustering out pay was $100 in cash plus $100 per month for the next two months. In short, Dad was ready to party. And from what I understand, party he did.

Dad and his boyhood friends from West Allis—"The Old Gang" or simply "The Gang,"—later to include wives. Top, left to right: Arnie Friedman ("Irish"); Rich Bernier; Jack Jung; and Paul Franson ("Pots"). Bottom: Don Andrews ("Andy"); John Hammer ("Lard"); Dad; and Max Shelby ("Boots").[c]

Claude and Anna Bernier, parents of his friend Rich Bernier, owned a tavern on Greenfield Avenue called "Frenchie's." A flat above the tavern was vacant, and Claude let Dad live there temporarily. All Dad had to do was make sure the stoker in the coal-fired furnace was full and remove the "clinkers" that were left after the coal burned.

His friend Rich Bernier had a cousin named Marge Norman, the same age as Dad, whom Dad had encountered a few times before the War. One night, Rich and Dad went to watch Marge bowl at a tavern on Greenfield Avenue which had a few bowling alleys in the basement. During the course of one game, Marge—much grown up since Dad last saw her—had a 7-10 split, the hardest split in bowling. Dad said, "I'll buy you a box of chocolates if you make that." Unbelievably, she made the split, Dad bought the chocolates, and Dad and Marge started dating. Most of Dad's veteran boyhood pals were also dating soon. The partying continued, sometimes just the guys and sometimes as couples.

Dad and Mom with Dad's green 1937 straight-8 Studebaker President 2-door Custom Coupe. He called it the "Green Hornet."[c]

Within a few years all were married. For the rest of their lives, Dad and his friends—and their wives—called themselves "The Gang."

According to the Selective Service and Training Act of 1940, an employer was supposed to rehire a draftee who asked for his old job back within forty days of being separated from active duty. I do not know if Dad made it within the forty days, but eventually he did go back to work at the A&P store across from City Hall on Greenfield Avenue.

By the time spring came, Dad and Marge—or should I call her "Mom"—were planning their wedding.

While Dad and Mom were dating, things were not going so well back in Italy for the monarchy that had backed Mussolini for so many years. Dad must have been amused to learn the Italian people, in a referendum on June 2, 1946, voted to end the monarchy and form a democratic republic.

Mom had grown up on the South Side of Milwaukee, near Greenfield Avenue, roughly forty blocks east of where Dad grew up in West Allis. She went to St. Matthew's Grade School and graduated from Mercy High School in 1943. She attended business school and went to work at International Harvester Company in Milwaukee during the War as a stenographer.

Mom and Dad's wedding on August 24, 1946. To the right of Dad is Uncle Bill with little Jill. To the left of Mom are her parents, Harry and Victoria Norman.[c]

Her father, Harry Norman, worked as a precision grinder at the Milwaukee Works of International Harvester Company. He tended bar on weekends and was an usher at church on Sundays. Mom had a married older brother named Bernard, who was a foreman in their Uncle Al's factory in West Allis. The company was a defense contractor, and Bernard, who had been a second lieutenant in the National Guard before the war, did not have to go on active duty.

Mom's father Harry and mother Victoria, or "Vicky," were originally from rural Manistee County Michigan, and both came from big families. Mom had a lot of aunts, uncles, and cousins. Her father had only three years of education, her mother four.

In short, despite her parents' near functional illiteracy, Mom came from a hardworking, simple, stable, nuclear family—something Dad had never known.

On August 24, 1946—a little over *eight months* after Dad came home from Italy—Dad and Mom were married at St. Matthew's Church on the South Side of Milwaukee.

They were both twenty-one.

EPILOGUE

Grandma and Grandpa—Dad called them "Ma" and "Pa"—had lived for many years in the upper flat of a duplex at 33rd and Madison Street on Milwaukee's South Side. Grandpa got a lot of overtime during the War, and Grandma was very thrifty. After the War, helped by a loan from Grandma's brother-in-law, Al Flicek, they were able to buy a small house in West Allis, on South 83rd Street. When Grandma and Grandpa moved, Mom and Dad took over their flat on the South Side. That is where they were living when I was born, ten months after the wedding—part of what was called the postwar "Baby Boom."

Dad's skill with numbers got him promoted to *field auditor* for A&P, and he was "on the road" a lot as he traveled to stores all over the Wisconsin-Upper Michigan region as part of a crew taking inventory. He and Mom moved into the basement of Grandma and Grandpa's new house in West Allis in 1948 for a while, and then to a downstairs flat in Milwaukee, two blocks away from their first flat, near 32nd and Scott Streets, owned by the D'Amico's, in-laws of Dad's friend and Mom's cousin, Rich Bernier.

Within six months after I was born, Milwaukee's WTMJ became the first commercial TV station to start broadcasting in Wisconsin, and Dad bought a TV with a small 10-inch screen to keep Mom entertained while he was doing auditing work away from home. Grandpa, living in West Allis, became a loyal customer of the Knippel-Selig Ford dealership, located at 71st and National Avenue, a few blocks away from where Dad had gone to high school. (The owner, Ben Selig, had a son named Bud, who, years, later brought Seattle's baseball team to Milwaukee, to become the *Brewers*—and who subsequently became the Commissioner of Major League Baseball).

My sister Mary was born in 1949. Dad, having become protégé of an A&P district supervisor, was admitted to a management training program; and in 1951, Dad's dream since his time in Foggia, if not before, came true. At age twenty-six, he was made *manager* of an A&P grocery store at the corner of Cedar and Jefferson Streets in downtown Green Bay, Wisconsin (population 53,000), home of the NFL "Packers" professional football team. What Dad anticipated would be a two to three year assignment in Green Bay turned into *nine years* when Dad's mentor died unexpectedly early in his tenure.

My mother learned to drive in Green Bay—Wisconsin's oldest city—where the French had established a trading post in 1634. My grade school, *Howe Elementary*, named for a U.S. Senator from Green Bay, actually began instruction in 1868; though the "new" redbrick building where I attended had replaced the original in 1913.

My sister Pat was born in Green Bay in 1952. Unfortunately, the birth involved life-threatening hemorrhaging for Mom; and shortly afterwards Mom was given a "partial hysterectomy with unilateral oophorectomy."

As his new family was growing, Dad's old family was getting smaller. A few months before my sister Pat was born, Dad's institutionalized mother Rosaline, forty-eight, died of a seizure during an epileptic bout. Dad was given her effects—a few pictures, an old letter from her husband (still alive at the time, but who knew?), a prayer book, etc.—in a small wooden box that had been used by her grandfather, William Lark, as a handkerchief box in the Civil War. Ray Blake, husband of Dad's cousin Kay, with whom Dad used to live, had a cardiac condition that had kept him out of service. Unfortunately, he died after open heart surgery in 1954 at age thirty-nine, leaving Kay a widow with three young children.

Dad with his ailing Uncle Bill. 1954.[c]

Eight months later, Uncle Bill died of colorectal cancer. After Bill's death, his "wife" Helene gave Dad $500. Before dying, Bill told Dad he and Helene said they had *married* on a trip to Arkansas several years before—but they had not really gone through with the formalities. Dad's mother and Uncle Bill were buried in the same cemetery lot with their mother, Dad's grandmother, Catherine Spearman, in Evanston, suburban Chicago, Illinois.

We Finney kids were generally happy in 1950s Green Bay, derisively called "Toilet Paper Capital of the World," home to Northern, Charmin, and Ft. Howard paper mills. The city had two rivers, the Fox and a tributary, the East River, which were among the most polluted waterways in America. Air quality was also poor. I remember my lungs used to burn before every rainfall due to sulfur dioxide and other air pollutants from the mills.

Dad and Mom rented a two-story house, built in 1909, at Crooks and Van Buren Streets on the East Side, in the older part of the city. Green Bay did not have television when we first moved there in 1951. When the Federal Communications Commission imposed a four-year moratorium on new licenses in 1948, the only TV station in Wisconsin was WTMJ in Milwaukee. In March of 1953, after the moratorium was lifted, WBAY in Green Bay became the second TV station in the state, and Dad bought an RCA mahogany 17-inch black-and-white console, which was our family television for many years.

In 1955, my parents bought their first house, on the GI Bill, at 708 Crooks Street, an old two-story, three-bedroom place two blocks away from where we had been living. That area of Green Bay was originally platted in 1835 as Astor Town, a development owned in part by nineteenth century New York fur magnate—and richest man in America—John Jacob Astor. Dad converted the asbestos-covered gravity furnace from coal to oil, remodeled a front-hall closet into a powder room, tore off the front porch, demolished the old garage, and built a recreation room with a small pool table in the basement.

In an era before bank cards and ATMs, and when banks closed at three in the afternoon, people who needed to cash personal and payroll checks often went to the check-cashing window at Dad's office in the A&P. Dad's store was downtown, not too far from City Hall, and he met a lot of prominent people in Green Bay by cashing their checks.

In June of 1957, A&P built a second grocery store in Green Bay on the west side at 1051 West Mason Street, and Dad was made manager of that store. The mayor was at the ribbon-cutting at the store's grand opening, which also featured baseball players Johnny Logan and Frank Torre (brother of Joe) of the Milwaukee Braves team that would go on later that year to win the 1957 World Series against the Yankees. The Braves had moved from Boston to Milwaukee in 1953. Dad's store was visited by Senator—later President—John F. Kennedy during the 1960 Wisconsin democratic primary campaign. In spring of that year, Dad and I attended a CYO sports breakfast where the guest of honor was the rookie head coach of the Green Bay Packers—who had produced the team's first winning season in twelve years—a fellow named Vince Lombardi.

In 1960, Dad was transferred back to the Milwaukee area to manage an A&P store on the North Side. The five of us moved to a small, brand-new, three-bedroom one-bath ranch house in West Allis on 85th Street. It had half the square footage of our old house in Green Bay, but was only two blocks from where Mom's parents were living. The house had no garage and the driveway was unpaved. Dad quickly had the driveway paved and built a two-and-a-half-car garage. Dad was happy to be back in West Allis, where he had grown up—and back with "The Gang." According to the 1960 census, West Allis was actually the fourth largest city in Wisconsin at the time, slightly bigger than Kenosha and Green Bay. The downside was a long commute. The new house was on the other side of town from the A&P store Dad was managing, which was in a high-crime area and had a shoplifting problem.

As time went on, Dad became more and more disenchanted with his employer, and more *vocal*. When the curtains parted at one management dinner to reveal how A&P was going to regain market share, he blurted out, "Not another dish promotion!"

Dad was not without his flaws. Although he fought to keep himself under control, the "street kid" in him came out from time to time. He could be combative and profane and had a violent temper—though the violence was never directed at Mom or at us. I recall his picking up a chair once when we lived in Green Bay and slamming it down on the dining room table, damaging the tabletop and a leg on the chair. Once, in West Allis, he put his fist through the back of our hollow front-closet door in a fit of anger. In another incident, he threw a plate of spaghetti. Like so many people, Dad and Mom lived paycheck to paycheck and were beset with financial problems. As a kid and during the War, Dad had done a lot of gambling. As money got tighter, he was again attracted to gambling. Of course, he lost, resulting in more heartache, an acknowledgement he had a problem—and with a determination not to gamble again.

The "mid-management squeeze" Dad was feeling at A&P in the 1960s, as his career there plateaued, caused a lot of stress. My parents tried to shield us children from the financial problems. We never knew Dad's salary, but we knew money was tight as we entered our teens.

Eventually, Dad missed making a mortgage payment or two. Foreclosure was threatened, and it looked like we were going to lose the house. Reluctantly—angry at Dad, and with tears in her eyes—Mom had to appeal to Grandma and Grandpa, who had been savers all their lives, and who came to the rescue. Mom brought in extra income waitressing at banquets with Grandma, and she demonstrated Kenner toys at department stores to earn extra money at Christmas.

The original "Yum-Yum Wagon."[c]

Nevertheless, Mom and Dad's relationship suffered a major upheaval. Mom's surgery after Pat's birth at age twenty-seven, with removal of an estrogen-producing ovary, had thrown Mom into a hormonal tailspin in her early thirties (premature ovarian failure). Dad was not the ladies' man his Uncle Bill was; but I could tell women were attracted to him. The upshot was Dad, who never wore a wedding ring, was accused of an extramarital affair in the early 1960s, leading to a great deal of discord in the household—and nearly ending in my parents getting divorced.

One day Mom got pale and passed out on the floor of our small kitchen. She was rushed to the hospital, was given a lot of tests, and received several pints of blood. Dad came home from the hospital to tell us three kids Mom was *bleeding internally*. Dad said Mom's severe hemorrhaging was life-threatening, and the doctors did not know what was causing it. As there were no MRIs or CT scans in those days, an "exploratory surgery" would be necessary to find the cause of the bleeding—and the doctors were not sure Mom was strong enough to undergo it. The doctors said her chances of pulling through were *less than fifty-fifty*. As Dad told us this, tears came to his eyes. Men were not supposed to cry in those days, and I had never seen him cry before. Facing the prospect of life without Mom, he broke down in tears right in front of us. I knew then how much he really loved Mom. She needed numerous transfusions, but the surgery was successful. The life-threatening problem, it turned out, was a bleeding duodenal ulcer.

The pressures, financial and otherwise, Dad was under in this unhappy

Dad in the second "Yum-Yum Wagon" on W. Greenfield Avenue, across from the Wisconsin State Fair Park.

time pushed him toward taking baby steps in the direction of what he would eventually become—an entrepreneur, of sorts.

One summer in the 1960s Dad converted an old Volkswagen minibus into a popcorn truck called the "Yum-Yum Wagon," which Mom and we kids operated.

The next summer, he bought an old laundry truck, which he adapted as a much bigger "Yum-Yum Wagon," eventually adding a generator and cotton candy machine. We also sold soft drinks, candy, and sno-cones. Mom and my sisters ran it for several summers, setting up at the West Allis Farmer's Market, at baseball games, parades, company picnics, etc.

. . .

In about 1967, with Mom's moral support, Dad resigned from A&P, where he once thought he would be employed for life. He saw little opportunity for advancement as the company that was once America's biggest retailer kept losing market share to upstart rivals who built beautiful new grocery stores with large parking lots. He sold insurance briefly and took

Dad at West Allis Pool & Supply in 1990.

classes in salesmanship, but missed the retail business and talked his way into a job managing a store that sold pool tables—despite disliking the owner's "bait-and-switch" tactics.

In 1968, with one pool table and a classified ad in the Milwaukee *Journal,* he started a pool table and billiard equipment establishment of his own, called *West Allis Pool & Supply,* in a leased storefront on Beloit Road at 64th Street, later moved to 69th and Beloit Road.

Since foundations of houses in the Milwaukee area had to be dug below the frost line, almost all the houses had basements. Dad's typical customer was someone looking to put a pool table in a basement recreation room. Among his competitors were the "big box" stores, called "discount department stores" back then.

The word "Pool" in the name of the business also referred to *swimming pools*; and Dad sold above-ground swimming pools and swimming pool chemicals in the summers, when the billiard business dropped off.

In about 1971, while still running the store, he leased warehouse space at 54th and Electric Avenue for a couple of years. The warehouse space was too big for his needs, so he bought a number of 9-ft pool tables and used part of the space to operate a *pool hall* called "The Corner

Part of "The Gang" in 1971. Top: Rich Bernier, Jack Jung, Dad. Bottom: Don Andrews and Paul Franson. (All are gone now. Their buddy John Hammer died of a heart attack in 1968 at age 44, the first to go).

Pocket," where I worked for several months before going to the UW in Madison for graduate school.

West Allis Pool & Supply relocated three times before Dad bought a store in 1977 that was in a multiuse property at 89th and National Ave. in West Allis. Grandma, a recent widow, lived in an apartment above the store, and my parents, "empty nesters" for several years by then, lived in the apartment adjacent.

In 1982, having hired his retired friend Rich Bernier, Dad started manufacturing a cue-tip scuffing and shaping device he called the "Cue Cube," eventually sold throughout the U.S and in a number of foreign countries, a business he expanded and continued to operate after health issues resulted in his selling West Allis Pool & Supply and going into semiretirement toward the end of his life.

Like many undercapitalized entrepreneurs, Dad had his share of cash-flow problems over the years. The old cliché "robbing Peter to pay Paul" about describes it. If I answered a phone call from a creditor while helping out at the store from time to time, I was instructed to say, "The check is in the mail." Nevertheless, Dad was a good salesman, and his business eventually prospered, especially when supplemented by sales and repairs of pool cues and recreation room accessories.

Cue Cube.

While operating his various businesses, Dad promoted the game of pool as best he could. Though not much of a pool player himself, he attended annual trade shows in Chicago, Las Vegas, and other cities, and sponsored exhibitions with celebrated pool players of the time.

Once, he had an event with the legendary Willie Mosconi—who played *golf* with Dad. Another time he sponsored "Minnesota Fats" (Rudolf Wanderone), who was actually more skillful as an entertainer than as a pool player.

Dad sponsored pool tours and for several years was president of the Billiard Congress of Wisconsin. He helped sponsor a television pool show on Channel 6 in Milwaukee hosted by noted Wisconsin sportscaster Earl Gillespie. He also advertised on a weekly Italian-language radio program in Milwaukee hosted by veteran broadcaster Giorgio Argondizzo. (I actually appeared on it once, speaking my very bad Italian).

With Dad's mess sergeant and grocery store background, he saw to it we always ate well. Even during the worst periods of my childhood, I remember we also had happy times. Dad always emphasized the importance of a sense of humor and liked to tell jokes. We were taught laughter and a sense of humor got you through hard times.

When Dad was a child, with no parents to speak of, he envied his friends on *family occasions*—especially Christmas. So, when he had a family of his own, Dad always made Christmas something special—much to the delight of us kids. Mom and Dad planned for Christmas months in

When legendary pool player Willie Mosconi from Philadelphia came to Milwaukee for an event sponsored by Dad, he said he wanted to play golf. So Dad played with him. (Dad was a businessman, not much of a pool player himself—and even less of a golfer.)

advance and probably spent more than they should have on toys and gifts for us. But Christmas was always a fun time in the Finney household.

The mess sergeant in Dad came out on holidays. While Mom did most of the cooking, Dad, at Thanksgiving and Christmas, always made our turkey with a rich, meat-heavy stuffing. At Easter, he baked a ham, with a tasty cherry sauce.

When Dad sold West Allis Pool & Supply, a few years before he died, he and Mom moved with Grandma to a house on S. 75th St. in West Allis that included a smaller keystone lot next-door with a large tree on it. Grandma fell at home, twice, breaking her hip each time and had to be put into a nursing home. She eventually died at age ninety-seven of Alzheimer's, only a few days before Dad.

All three of Dad and Mom's children, by now adults, had built their

After Christmas Dad. (1950's Green Bay).[c]

Google Maps Street View of the house Dad built on S. 75th St. in West Allis in 1994. (©2017 *Google*).

own homes. Not to be outdone, and ever the entrepreneur, even in ill health, Dad had the tree cut down on the keystone lot and built a brand new three-bedroom ranch house in 1994, where he and Mom were supposed to live out their retirement years. Unfortunately, by the time it was finished, Dad only had about a year left.

The smoking habit Dad started in his youth and continued through his time in service and afterwards caught up with him. When he was sixty, in 1985, he suffered a heart attack and underwent cardiac bypass surgery at the VA. Unlike his father, grandfather, and great-grandfather, Dad never lived at the Veterans Administration home in Milwaukee; but he took advantage of the hospital there until he was eligible for Medicare. In the years following his heart attack, his heart, respiratory, circulatory, and kidney issues—among others—caused more and more frequent hospitalizations.

His ailments included emphysema, chronic bronchitis (with "smoker's cough"), coronary artery disease, heart failure, chronic urinary tract infections, multiple kidney stones, loss of a kidney, renal artery aneurism, inoperable hernia, intestinal problems, etc.

A trip to Italy, which he had not seen since the War, was on his "bucket list," however; and, as mentioned earlier, he visited for a week in March of 1994—able to walk short distances despite his many health problems.

A little over a year later, in May of 1995, fifty years after the War in Europe ended, the day after Grandma's funeral—which he was too ill to attend—in a hospital room surrounded by his family, Dad died of chronic obstructive pulmonary disease at age seventy, a smoker to the end.

Mom went into mourning, lost a lot of weight, and was quite distraught for a few years until she met a wonderful lifelong bachelor, Ron Marten, who helped her through her grief. They were married on Mom's seventy-third birthday.

Dad left behind children and grandchildren, and he now has thirty-one great-grandchildren (and counting).

FINAL THOUGHTS

Long gone are the War, Dad, "The Gang"—and West Allis.

Oh, the City of West Allis is still here, of course, but not as the place where Dad grew up. Gone are Dad's high school, the A&P, the orange-and-cream streetcars, the movie theaters with their bright marquees, and the bustling shoppers on Greenfield Avenue.

Gone also is the Allis-Chalmers mega-factory that gave birth to the community and where tens of thousands of workers were kept so busy during Wartime.

World War II, of course, was a terrible *world-wide* event that, in one way or another, affected the lives of literally hundreds of millions of people, great and small. The war planners in Washington and London had limited resources and had to fight a determined enemy. Priorities were established based on very practical *geographic* realities. The Atlantic Ocean being smaller than the Pacific, and the European enemy therefore more of a threat than Japan to the US and Britain—and with shorter supply lines—the Atlantic war took precedence over the Pacific war (much to Hitler's disappointment). Britain was close to the continent and was a place where an invasion force for Northern Europe could be assembled. Unlike Italy, Northern Europe was not separated from Germany by a formidable mountain range like the Alps. So, the European Theater became more important than the Mediterranean Theater, Dad's theater, the so-called "forgotten front."

While Dad's part of the War may have had to play second fiddle (or third or fourth) to the other fronts, the fighting in Italy was certainly as ferocious as elsewhere, fought with the same modern weapons, under extreme conditions—and the soldiers, sailors, and airmen equally brave.

Though the front may be "forgotten," we cannot forget those who fought there.

After the War, German general von Senger und Etterlin, who surrendered to General Mark Clark on behalf of Army Group C, offered this advice to English historian Sir Michael Howard: "The next time you invade Italy—*do not start at the bottom*!"

Italy saw some of the fiercest combat of the war under mountain conditions that, until the last few weeks of fighting, frustrated the use of large numbers of troops and modern mobile warfare. Much of the mountain fighting Dad participated in during the winter in 1945 involved units of, at most, regiment size, and usually smaller. Fifth and Eighth Armies in Italy were given the task of attacking enemy forces of nearly their own strength, when traditional dogma taught an attacking force should outnumber an entrenched defending force by *several times*—though the awesome Allied air power made victory only a matter of time.

The Allied armies in Italy represented many nations—a situation that required coordination of soldiers from different races and cultures, speaking many different languages. General Mark W. Clark, 15th Army Group commander, had to be a diplomat, as well as a soldier. His Allied troops in Italy had to make do with manpower and equipment not wanted or needed by the other fronts—all the while pinning down some of Germany's best units, so as to make them unavailable to fight elsewhere.

During the great worldwide Depression of the 1930s, Nazi military spending had seemed like the answer to terrible unemployment and brought Germany temporary economic prosperity. The military build-up of the 1930s, however, was financed by *deficit spending*. The Western powers (and the USSR) did not understand the corrupt Nazi political regime, which always knew its debts could only be managed by using its new military assets aggressively against its neighbors, especially to the East. The Nazi economy was always intended to be supported by the *theft* of natural resources and exploitation of what amounted to slave labor in the countries it conquered. Stripped of its colonies in Africa and the South Pacific as an outcome of World War I, Germany under the Nazis planned to "colonize" and exploit its European neighbors.

Yet, however rich in resources Germany was, however talented its scientists and engineers, however skillful its military tacticians, it was never motorized enough and never had sufficient natural or human resources to fight a prolonged two-front—much less a three-front—war.

The *Luftwaffe*, instrumental in Germany's early victories, ultimately lacked the fighters and reliable heavy bombers to win the Battle of Britain. Without air superiority, the *Luftwaffe's* Stuka dive bombers, so necessary for *Blitzkrieg* tactics, were easy pickings for Allied fighter planes. In 1944, the Allies manufactured four aircraft for every one Germany made. Germany also fell behind the Allies in maintenance and repair of aircraft and especially—due to limited fuel—in the training of skilled pilots. When the Allies achieved air superiority, disrupting German supply lines, the Nazis never had enough spare parts, transport planes, trucks, or fuel to keep their war machine going like earlier in the War.

Germany's defeats at Stalingrad and in Tunisia in 1943 turned the tide of the War. The Soviet Union's robust industrial production, its competent tanks, aircraft, and artillery, and its seemingly endless supply of manpower (not to mention Allied aid, and *the weather*) proved too much for the German military—especially with Hitler's meddling, overriding the better judgment of his generals.

Unlike World War I, which was largely fought outside of Germany, strategic bombing and invasion brought World War II to the average German's doorstep.

The defeat of Germany in World War II was utter, on all fronts.

Obviously, Dad, a mess sergeant—one of sixteen million Americans who served over the course of the War—was only a very small cog in a very big machine. Yet he was rightfully proud of his contribution. For about a year, he participated with Fifth Army in the Italian campaign, which resulted, after much difficulty, in the resounding defeat of determined German forces—a victory that cost Fifth Army, over the twenty months of the entire campaign, about twenty thousand killed; eighty thousand wounded; and ten thousand missing. Fifth Army was only as good as the sum of its parts and needed the contributions of its individual soldiers, like Dad.

Knowing Dad's personality, I am sure Dad, young as he was, was more than a good manager and leader during his time in service. He may have grown up on the streets, but he learned how to be a *charmer* when he wanted. While he was resourceful, responsible, and conscientious, he was also always ready with a comment, or joke, or story. Dad was a "people person," very approachable and personable, with many friends. He genuinely liked others and was liked by them. I am sure he was popular with his fellow soldiers.

The mess, after all, was a *communal* activity. Everyone had to be fed, high and low. The lower enlisted men were not only fed by Dad, they also worked for him when placed on detail—to help his cooks in the kitchen, or help serve food, or help clean up. I am sure everyone in the battery, from the Captain on down, knew Dad, and liked him.

When Dad worked for A&P in West Allis during high school and after graduation, he was not only a clerk but also created window and floor displays. He was a creative person and enjoyed putting things in the store windows and at the end of aisles to attract customers. Later, as a grocery store manager and when he had his own businesses, he enjoyed advertising and making his stores as attractive to the public as possible. I am sure, when in service, Dad's creativity came out in his mess. Centerpieces on the table, fresh fruit when available, special desserts or special treats of one kind or another, like ice cream, fresh local food, if he could procure it—these are things I am sure he provided.

Having served myself during the Vietnam War, I know few things are as important in the daily life of a soldier as his mess. Dad, the friendly mess sergeant, was not only responsible for the nutrition, of the hundred, or so, men in his battery. I know he contributed mightily to that elusive, intangible, yet vitally important quality of a functioning combat unit—*morale*.

Even though he was in a support role and earned no medals for heroism, Dad lived with fear, and was subject to enemy air attacks, sniper fire, and incoming artillery and mortar fire on numerous occasions. He came close to serious injury or death when he was asleep one night, and a mortar round landed by his foxhole and buried him alive, leaving him with nightmares for the rest of his life.

Working closely with the incredible 10th Mountain Division troops, Dad saw many dead and wounded from both sides, and among civilians—revolting sights that could not be erased from his memory, enough gore and human misery to last a lifetime. Like almost every combat veteran in World War II who returned home, he had experiences he lived with the rest of his life. He learned to "compartmentalize" so he could survive and move on. Some experiences he never shared with me or anyone else—and took with him to his grave.

Dad's story is only one of millions, of course, in most ways typical, in others unique.

I hope, however, we never forget those whose stories—unlike Dad's—*ended* tragically in Italy, and at the other fronts, with the *ultimate* sacrifice, and those who returned and had to adjust to life with permanent physical and emotional injuries. And we cannot forget the spouses, children, parents, siblings, and friends impacted by the death, wounding, or illness of a loved one who served.

As for Dad, *what would he have done if World War II had not come along?* I can only speculate. But afterwards, he had a sense of responsibility, along with leadership skills, resourcefulness, confidence in himself, and generally positive values he otherwise might not have had—especially, given the unfortunate circumstances into which he was born. Unlike him—but thanks to him—I was fortunate to have been born into a solid family unit, which provided happy childhood memories, like being overcome with joy and laughter when he became like a kid again and roughhoused with us on the living room carpet, or like on our long drives through snowy farm country between Green Bay and Milwaukee, when the five of us, led by Dad, sang Christmas carols together in our second-hand Cadillac. Our family had many happy moments.

Of course, fathers and sons do not always see "eye-to-eye." Dad smoked. I didn't. He drank hard liquor, but did not drink beer. I loved beer. He married at twenty-one. I married at thirty-three. After getting out of service, he almost never traveled. When I got back from Vietnam, I couldn't wait to explore Europe. He did not take advantage of GI Bill education benefits. I totally exhausted mine. Nevertheless, he mostly encouraged me in my life choices. No son could ask for more. I have fond memories of him—the smell of cigarettes always about him, the scent of his aftershave, and the roughness of his unshaven cheek when I kissed him as an old man. Dad, a store manager and former sergeant in the Army, instilled in me lessons of leadership that have guided me my whole life.

Thankfully, Dad was among those lucky enough to come home intact and be part of the Greatest Generation, to have a productive and full life in the postwar world, to experience love and marriage and have a family, and to pass on to his children his sense of humor, love of laughter and fun, and talent for enjoying life.

Possibly it was because of his experiences during the War, or his difficulties growing up without parents during the Great Depression, or both; but, as my sisters will attest, Dad always let us children know—we meant *everything* to him.

A Depression Era street-kid, a real product of his time, with admitted shortcomings, of course, but an endearing personality, Dad may have had almost nothing by way of family growing up; but when he passed away, he left behind a family that loved him dearly—*and still misses him.*

FOR FURTHER READING

FOR FURTHER READING

Alexander of Tunis, Field Marshal Earl. *The Alexander Memoirs, 1940-1945*. (Havertown, PA: Frontline Books, 2020).

Allison, Richard. *Begging for Chocolates: A Story of World War II Italy*. (Ashland, OR: Hellgate Press, 2017).

Ambrose, Stephen E. *Eisenhower and Berlin: The Decision to Halt at the Elbe*. (New York, NY: W.W. Norton, 1967/2000).

Armstrong, Frank H. *Payoff Artillery—WWII*. (Burlington, VT: Bull Run of Vermont, 1993).

Baumgardner, Randy. *Tenth Mountain Division Volume II*. (Paducah, KY: Turner Publ. Co., 2003).

Blackwell, Ian. *Fifth Army in Italy 1943-1945: A Coalition at War*. (Barnsley, UK: Pen & Sword Books, 2012).

Bosworth, R.J.B. *Claretta: Mussolini's Last Lover*. (New Haven, CT: Yale Univ. Press, 2017).
Brooks, Thomas R. *The War North of Rome: June 1944-May 1945*. (Edison, NJ: Castle Books, 1996).

Brownell, Will and Denise Drace-Brownell. *The First Nazi: Erich Ludendorff The Man Who Made Hitler Possible*. (London, UK: Duckworth Overlook, 2016).

Burbach, Rev. Julius H. *West Allis, A City of Marvelous Growth In A Decade, June 28, 1902-June 28, 1912*. (Milwaukee, WI: 1912).

Clark, General Mark W. Calculated Risk. (New York, NY: Enigma, 1950/2007).

Coltro, Gabriele. *Il Tesoro del Duce; La vera storia dei valori sequestrati a Dongo alla colonna Mussolini e finiti al Partito Communista.* (Florence, Italy: goWare, 2018).

Cooke, James J. *Chewing Gum, Candy Bars, and Beer: The Army PX in World War II.* (Columbia, MO: Univ. of Missouri Press, 2009).

Corvaja, Santi, translated by Robert L. Miller. *Hitler and Mussolini: The Secret Meetings.* (New York, NY: Enigma, 2008).

Deighton, Len. *Blood, Tears, and Folly: An Objective Look at World War II.* (London, UK: William Collins, 2014).

Delle Stelle, Pier Luigi Bellini and Urbano Lazzaro. Translated from the Italian by W. C. Darwell. *Dongo the Last Act.* (London, UK: MacDonald & Co., 1964).

Dollmann, Eugen. Translated from the German by J. Maxwell Brownjohn. *With Hitler and Mussolini: Memoirs of a Nazi Interpreter.* (New York, NY: Skyhorse, 1976/2017).

Dulles, Allen. *The Secret Surrender.* (New York, NY: Harper & Row, 1966).

Edsel, Robert M. *Saving Italy: The Race to Rescue a Nation's Treasures from the Nazis.* (New York, NY: W.W. Norton, 2013).

Ellis, Robert B. *See Naples and Die: A World War II Memoir.* (Jefferson, NC: McFarland & Co., 1996).

Ferguson, Harvey. *The Last Cavalryman: The Life of General Lucian K. Truscott, Jr.* (Norman, OK: Univ. of Oklahoma Press, 2015).

Feuer, A.B. *Packs On! Memoirs of the 10th Mountain Division in WWII.* (Mechanicsburg, PA: Stackpole Books, 2006).

Fisher, Ernest F., Jr. *Cassino to the Alps*. (London, UK: Forgotten Books, 2015 reprint from (Washington, DC: GPO, 1977).

Fisher, John C. and Carol Fisher. *Food in the American Military: A History*. (Jefferson, NC: McFarland & Co., 2011).

Fifth Army /Headquarters IV Corps, U.S. Army. Nance, Col. Curtis H. ed. *19 Days from the Apennines to the Alps, The Story of the Po Valley Campaign/ The Final Campaign across Northwest Italy 14 April -2 May 1945*. (Nashville, TN: Battery Press, 1987; reprints from Milan, Italy: Pizi and Pizio, 1945).

Goodspeed, D.J. *Ludendorf: Soldier: Dictator: Revolutionary*. (London, UK: Hart-Davis, 1966).

Gowen, Kenneth K. *Granddaddy, Tell Us About the War: A Southern GI's Experiences in WWII*. (Oxford, MS: Kay-Dot Publ., 1998).

Hamann, Brigitte, translated by Thomas Thurnton. *Hitler's Vienna: A Dictator's Apprenticeship*. (Oxford, UK: Oxford Univ. Press, 1999).

Hanfstängl, Ernst, translated by John W. Toland. *Hitler: The Memoir of a Nazi Insider Who Turned Against the Führer*. (New York, NY: Arcade, 2011).

Hibbert, Christopher. *Benito Mussolini*. (London, UK: Penguin, 1965).

Holland, James. *Italy's Sorrow: A Year of War, 1944-1945*. (New York, NY: St. Martin's Press, 2008).

Jennings, Christian. *At War on the Gothic Line: Fighting In Italy, 1944-45*, (New York, NY: St. Martin's Press, 2016).

Jordan, Jonathon W. *American Warlords: How Roosevelt's High Command Led America to Victory in WWII*. (New York, NY: Caliber, 2015).

Junge, Traudl. Translated from the German by Anthea Bell. *Hitler's Last*

Secretary: A Firsthand Account of Life with Hitler. (New York, NY: Arcade, 2003/2011).

Kershaw, Ian. *Hitler 1889-1936: Hubris.* (New York, NY: W.W. Norton, 1999).

Kershaw, Ian. *Hitler1936-45: Nemesis.* (New York, NY: W.W. Norton, 2000).

Kershaw, Ian. *To Hell and Back: Europe, 1914-1949.* (New York, NY: Viking, 2015).

Kesselring, Albert. English translation of *Soldat bis zum Letzten Tag* (1953). *Kesselring: A Soldier's Record* (New York, NY: Wm. Morrow & Co., 1954); (ebook) *The Memoirs of Field-Marshal Kesselring* (New York, NY: Skyhorse, 2016).

Klemperer, Victor, translated by Jessica Spengler. *Munich 1919: Diary of a Revolution.* (Malden, MA: Polity, 2017).

Lanfranchi, Ferruccio. *La Resa degli Ottocentomila: Con le Memorie Autografe del Barone Luigi Parilli.* (Milan, Italy: Rizzoli, 1948).

Lazzero, Ricciotti. *Il Sacco d'Italia: Razze e Stragi Tedesche nella Repubblica di Salò.* (Milan Italy: A. Mondadori, 1994).

Leahy, Fleet Admiral William D. *I Was There: The Personal Story of the Chief of Staff to Presidents Roosevelt and Truman Based on His Notes and Diaries Made at the Time.* (New York, NY:Whittlesey House, 1950).

Lewis, Norman. *Naples '44: A World War II Diary of Occupied Italy.* (New York, NY: Carroll & Graf, 1978).

Linge, Heinz, translated by Geoffrey Brooks. *With Hitler to the End: The Memoirs of Adolf Hitler's Valet.* (New York, NY: Frontline, 2013).

Lynch, Patricia A. *Milwaukee's Soldiers Home.* (Charleston, SC: Arcadia, 2013).

Mastalli, Pierfranco. *L'Arresto di Mussolini a Dongo e la Resa della Colonna Tedesca a Morbegno (27 e 28 Aprile 1945).* (Lecco, Italy: Cattaneo, 2012).

Meinke, Albert H., Jr., M.D. *Mountain Troops and Medics: Wartime Stories of a Frontline Surgeon in the U.S. Ski Troops.* (Kewadin, MI: Rucksack Publ., 1993).

Moretti, Rino. *Cattura e Morte di un Dittatore.* (Milan, Italy: Mursia, 2018).

Morris, Sylvia Jukes. *Price of Fame: The Honorable Clare Boothe Luce.* (New York, NY: Random House, 2014).

Moseley, Ray. *Mussolini, The Last 600 Days of Il Duce* (Lanham, MD: Lyons Press, 2004).

Muchka, Albert. *West Allis.* (Charleston, SC: Arcadia, 2003).

Mussolini, Benito. *My Autobiography,* translated from the Italian by Richard Washburn Child. (New York, NY: Charles Scribner's Sons, 1928).

Mussolini, Rachele, as told to Albert Zarca. *Mussolini: An Intimate Biography By His Widow.* (New York, NY: William Morrow, 1974).

O'Brien, Phillips Payson. *The Second Most Powerful Man in the World: The Life of Admiral William D. Leahy, Roosevelt's Chief of Staff.* (New York, NY: Dutton, 2019).

Payne, Robert. *The Life and Death of Adolf Hitler.* (New York, NY: Popular Library, 1995).

Perretta, Giusto. *La Verità: Dongo 28 Aprile 1945: Il Racconto di Michele Moretti.* (Como, Italy: Editrice Lariologo, 2018).

Peterson, Walter F. *An Industrial Heritage: Allis-Chalmers Corporation.* (Milwaukee, WI: Milwaukee County Historical Society, 1976).

Roberts, Andrew. *The Storm of War: A New History of the Second World War.* (New York, NY: Harper, 2011).

Ryback, Timothy W. *Hitler's Private Library: The Books That Shaped His Life.* (New York, NY: Alfred A. Knopf, 2008).

Schroeder, Christa, translated from the German by Geoffrey Brooks. *He Was*

My Chief: The Memoirs of Adolf Hitler's Secretary. (South Yorkshire, UK: Pen & Sword, 2009).

Schwarzwäller, Wulf. *The Unknown Hitler, Behind the Image of History's Darkest Name,* translated from the German by Aurelius von Kappau. (New York, NY: Berkley, 1990).

Shirer, William L. *The Rise and Fall of the Third Reich*. (London, UK: Secker & Warburg, 1960).

Speer, Albert, translated from the German by Richard and Clara Winston. *Inside the Third Reich*. (New York, NY: Avon, 1970).

Spinosa, Antonio. *I Figli del Duce*. (Milan, Italy: Rizzoli, 1983).

Starr, Lt. Col. Chester G., et al., Historical Section, Fifth Army. *Fifth Army History Part VIII—The Second Winter*. (Washington, D.C.: GPO, 1947).

Starr, Lt. Col. Chester G., et al., Historical Section, Fifth Army. *Fifth Army History Part IX—Race to the Alps*. (Washington, D.C.: GPO, 1947).

Toland, John. *The Last 100 Days*. (New York, NY: Random House 1965).

Truscott, Lt. General Lucian K. Truscott, Jr. *Command Missions: A Personal Story*. (Cleveland, OH: Pickle Partners Publ., 1954/2013).
Tyndall, Clifford. *Greetings from Camp Davis: The History of a WWII Army Base*. (Chapel Hill, NC: Chapel Hill Press, 2006).

Von Lang, Jochen. Translated from the German by MaryBeth Friedrich. *Top Nazi: SS General Karl Wolff, The Man Between Hitler and Himmler*. (New York, NY: Enigma, 2005).

Von Senger Und Etterlin, General Frido. Translated from the German by George Malcom. *Neither Fear Nor Hope*. (Novato, CA: Presidio Press, 1989).

Warner, Leonard. *Metamorphosis in the Mediterranean.* (Bloomington, IN: Xlibris, 2006).

Weber, Thomas. *Hitler's First War: Adolf Hitler, The Men of the List Regiment, and the First World War.* (Oxford, UK: Oxford Univ. Press, 2010).

Weber, Thomas. *Becoming Hitler: The Making of a Nazi.* (New York, NY: Basic Books, 2017).

Wilhelm, Maria de Blasio. *The Other Italy: The Italian Resistance in World War II.* (New York, NY: W.W. Norton, 1988).

Winterbotham, F.W. *The Ultra Secret.* (New York, NY: Harper & Row, 1974).
Wylie, Neville. *Britain, Switzerland, and the Second World War.* (Oxford, UK: Oxford Univ. Press, 2003).

Zaloga, Steven J. *M7 Priest 105mm Howitzer Motor Carriage.* (Oxford, UK: Osprey Publ., 2013).

Zaloga, Steven J. *US Field Artillery of World War II.* (Oxford, UK: Osprey, 2007).

Zoller, Albert. *Douze Ans Auprès de Hitler.* (Paris, France: Arthème Fayard, 1949).

INDEX

1

2

3

4

C

G

M

N

O

P

Q

R

S

X

Y

Z

ACKNOWLEDGEMENTS

I wish to thank the National Personnel Records Center and the National Archives and Records Administration (NARA) for their work in reconstructing my father's military records after the disastrous fire of 1973. Without their efforts, my father's story could never have been told. I also wish to thank the NPRC for their help when I visited their facility in St. Louis, Missouri to review morning reports of Dad's units. Thanks, too, to the Dwight D. Eisenhower Presidential Library in Abilene, Kansas for furnishing me with information on the 1125th Armored Field Artillery Battalion. I also wish to thank the West Allis Historical Society and the West Allis Public Library for their help. The Wisconsin Veterans Museum in Madison, Wisconsin has also been most helpful. The Denver Public Library Western History Department has become the repository of documents related to the 10th Mountain Division, which originally trained at Camp Hale in Colorado. I thank them for their help in obtaining photographs of the division and its operations in World War II. Thanks, too, to the *Museo della Fine della Guerra* in Dongo, Italy, which I visited, and the National World War II Museum in New Orleans, which I encourage everyone to see.

ABOUT THE AUTHOR

William P. Finney, a retired attorney, is a native Milwaukeean who grew up in Green Bay and West Allis, Wisconsin. A graduate of the University of Wisconsin—Milwaukee, he has a master's degree in Communication Arts and a law degree from the University of Wisconsin—Madison. He also holds a certificate in Italian from the *Università Italiana per Stranieri* in Perugia, Italy. Bill served in the U.S. Army in Phu Bai, Vietnam as a clerk-typist for the 101st Airborne Division. He met his wife Cheryl while in law school in Madison. Before moving to Los Angeles with Cheryl, he worked as a radio and television newscaster in the Madison area. Bill practiced law at a private firm in downtown Los Angeles, and Cheryl worked as an attorney in the insurance industry. After raising their three children, in Santa Monica, California, Bill and Cheryl lived in Hong Kong for several years and then moved to the Tampa Bay area of Florida.

www.ingramcontent.com/pod-product-compliance
Ingram Content Group UK Ltd.
Pitfield, Milton Keynes, MK11 3LW, UK
UKRC030757141026
15625UKWH00084B/690/J